A Review of the Landscape Conservation Cooperatives

Committee on the Evaluation of the Landscape Conservation Cooperatives

Board on Atmospheric Sciences and Climate

Board on Agriculture and Natural Resources

Division on Earth and Life Studies

The National Academies of
SCIENCES · ENGINEERING · MEDICINE

THE NATIONAL ACADEMIES PRESS
Washington, DC
www.nap.edu

THE NATIONAL ACADEMIES PRESS 500 Fifth Street, NW Washington, DC 20001

This activity was supported by Contract No. F14PC00112 from the U.S. Fish and Wildlife Service. Any opinions, findings, conclusions, or recommendations expressed in this publication do not necessarily reflect the views of any organization or agency that provided support for the project.

International Standard Book Number-13: 978-0-309-37985-4
International Standard Book Number-10: 0-309-37985-7
Digital Object Identifier: 10.17226/21829

Additional copies of this report are available for sale from the National Academies Press, 500 Fifth Street, NW, Keck 360, Washington, DC 20001; (800) 624-6242 or (202) 334-3313; http://www.nap.edu.

Printed in the United States of America

Suggested citation: National Academies of Sciences, Engineering, and Medicine. 2016. *A Review of the Landscape Conservation Cooperatives*. Washington, DC: The National Academies Press. doi: 10.17226/21829.

The National Academies of

SCIENCES · ENGINEERING · MEDICINE

The **National Academy of Sciences** was established in 1863 by an Act of Congress, signed by President Lincoln, as a private, nongovernmental institution to advise the nation on issues related to science and technology. Members are elected by their peers for outstanding contributions to research. Dr. Ralph J. Cicerone is president.

The **National Academy of Engineering** was established in 1964 under the charter of the National Academy of Sciences to bring the practices of engineering to advising the nation. Members are elected by their peers for extraordinary contributions to engineering. Dr. C. D. Mote, Jr., is president.

The **National Academy of Medicine** (formerly the Institute of Medicine) was established in 1970 under the charter of the National Academy of Sciences to advise the nation on medical and health issues. Members are elected by their peers for distinguished contributions to medicine and health. Dr. Victor J. Dzau is president.

The three Academies work together as the **National Academies of Sciences, Engineering, and Medicine** to provide independent, objective analysis and advice to the nation and conduct other activities to solve complex problems and inform public policy decisions. The Academies also encourage education and research, recognize outstanding contributions to knowledge, and increase public understanding in matters of science, engineering, and medicine.

Learn more about the National Academies of Sciences, Engineering, and Medicine at **www.national-academies.org**.

COMMITTEE ON THE EVALUATION OF THE LANDSCAPE CONSERVATION COOPERATIVES

DOROTHY J. MERRITTS (*Chair*), Franklin and Marshall College, Lancaster, Pennsylvania
BRENDA BARRETT, Pennsylvania Department of Conservation and Natural Resources (retired)
F. STUART CHAPIN, University of Alaska, Fairbanks
HOLLY D. DOREMUS, University of California, Berkeley
CRAIG GROVES, The Nature Conservancy, Bozeman, Montana
KENNETH HADDAD, Florida Fish and Wildlife Conservation Commission (retired)
JESSICA HELLMANN, University of Minnesota, Saint Paul
LYNN MAGUIRE, Duke University, Durham, North Carolina
PHILIP W. MOTE, Oregon State University, Corvallis
JOHN A. O'LEARY, Massachusetts Division of Fisheries and Wildlife (retired)
REBECCA R. RUBIN, Marstel-Day, LLC, Fredericksburg, Virginia
DALE STRICKLAND, Western EcoSystems Technology, Inc., Cheyenne, Wyoming
ERIC TOMAN, The Ohio State University, Columbus

Staff

CLAUDIA MENGELT, Study Director, Ocean Studies Board
DAVID POLICANSKY, Scholar, Board on Environmental Studies and Toxicology
STACEE KARRAS, Associate Program Officer, Ocean Studies Board
HEATHER COLEMAN, Postdoctoral Fellow, Ocean Studies Board
JENNA BRISCOE, Senior Program Assistant, Board on Agriculture and Natural Resources

BOARD ON ATMOSPHERIC SCIENCES AND CLIMATE

A.R. RAVISHANKARA (*Chair*), Colorado State University, Fort Collins
GERALD A. MEEHL (*Vice Chair*), National Center for Atmospheric Research, Boulder, Colorado
LANCE F. BOSART, University at Albany, State University of New York
MARK A. CANE, Columbia University, Palisades, New York
SHUYI S. CHEN, University of Miami, Florida
HEIDI CULLEN, Climate Central, Princeton, New Jersey
PAMELA EMCH, Northrop Grumman Aerospace Systems, Redondo Beach, California
ARLENE FIORE, Columbia University, Palisades, New York
WILLIAM B. GAIL, Global Weather Corporation, Boulder, Colorado
LISA GODDARD, Columbia University, Palisades, New York
MAURA HAGAN, Utah State University, Logan
TERRI S. HOGUE, Colorado School of Mines, Golden
ANTHONY JANETOS, Boston University, Massachusetts
EVERETTE JOSEPH, University at Albany, State University of New York
RONALD "NICK" KEENER, JR., Duke Energy Corporation, Charlotte, North Carolina
JOHN R. NORDGREN, The Climate Resilience Fund, Bainbridge Island, Washington
JONATHAN OVERPECK, University of Arizona, Tucson
ARISTIDES A.N. PATRINOS, New York University, Brooklyn
S.T. RAO, North Carolina State University, Raleigh
DAVID A. ROBINSON, Rutgers, The State University of New Jersey, Piscataway
CLAUDIA TEBALDI, National Center for Atmospheric Research, Climate Central, Boulder, Colorado

Ocean Studies Board Liaison

DAVID HALPERN, Jet Propulsion Laboratory, Pasadena, California

Polar Research Board Liaison

JENNIFER FRANCIS, Rutgers, The State University of New Jersey, Marion, Massachusetts

Staff

AMANDA STAUDT, Director
EDWARD DUNLEA, Senior Program Officer
LAURIE GELLER, Program Director
KATHERINE THOMAS, Senior Program Officer
LAUREN EVERETT, Program Officer
ALISON MACALADY, Program Officer
AMANDA PURCELL, Associate Program Officer
RITA GASKINS, Administrative Coordinator
ROB GREENWAY, Program Associate
SHELLY FREELAND, Financial Associate
MICHAEL HUDSON, Senior Program Assistant
ERIN MARKOVICH, Program Assistant

Preface

In 2014, the National Research Council convened a committee to examine the Landscape Conservation Cooperatives (LCC) program, drawing upon members whose expertise spans atmospheric sciences, biology, ecology, forestry, marine sciences, plant physiology, zoology, geomorphology, environmental policy, resource management, and decision analysis. This span of expertise reflects the need to encompass a view as wide ranging as that of a landscape-scale approach to conservation. A landscape-scale approach significantly broadens the 20th-century practice of conservation. In this report, "a landscape is defined as a large area encompassing an interacting mosaic of ecosystems and human systems that is characterized by a set of intersecting management concerns. The landscape is not defined by the size of the area, but rather by the interacting elements that are meaningful to the management objectives."[1]

The essence of conservation is to preserve, guard, protect, and use wisely. Commonly the word conservation is associated with wildlife, soil, water, and habitat, each of them a resource considered important to manage. Landscape conservation encompasses all of these and more, extending to the interacting mosaic of ecosystems and human systems and the many drivers that impact them. Conservation managers recognized the need for a landscape-scale approach, and in 2009, the U.S. Department of the Interior's Secretary Ken Salazar issued Secretarial Order No. 3289 to establish the LCCs. Broadening the scope of conservation to the scale of landscapes also requires cooperation and partnerships among many entities that themselves span institutional and geographic boundaries.

With this background, the committee's task was to examine the LCC program 5 years after it was established. The breadth of expertise of committee members was invaluable as it considered a landscape-scale approach to conservation, met with resource managers and policy makers representing a range of jurisdictional scales, and assessed impacts of the LCC program on the health of fish, wildlife, and their habitats. Many of the issues required greater understanding of how partnerships work across jurisdictional boundaries, and the direct experience of members from their own work was crucial. I am grateful to the committee members for their insights, thoughtfulness, and ability to develop consensus during this process. The committee is appreciative of the time and responsiveness of many who met with us and responded to our requests for more information, and their help enhanced our work.

I give special thanks and admiration to the National Academies staff—Claudia Mengelt, Study Director; David Policanksy, Scholar; Stacee Karras, Research Associate; and Jenna Briscoe, Senior Program Assistant—for their keen insights, thorough attention to detail, and dedication to this effort. It was a pleasure to work with such a professional group.

Dorothy Merritts, *Chair*

[1] Based on definition of "landscape scale" from the President's Priority Agenda: *Enhancing the Climate Resilience of America's Natural Resources*, which can be found at https://www.whitehouse.gov/sites/default/files/docs/enhancing_climate_resilience_of_americas_natural_resources.pdf.

Acknowledgments

This report has been reviewed in draft form by persons chosen for their diverse perspectives and technical expertise. The purpose of this independent review is to provide candid and critical comments to assist the institution in making its published report as sound as possible and to ensure that the report meets institutional standards of objectivity, evidence, and responsiveness to the study charge. The review comments and draft manuscript remain confidential to protect the integrity of the deliberative process. We wish to thank the following for their review of this report:

Douglas Austen, American Fisheries Society, Bethesda, Maryland
John Frampton, Council to Advance Hunting and the Shooting Sports, Washington, DC
William Graf, University of South Carolina, Columbia
Merlin Hanauer, Sonoma State University, Rohnert Park, California
Terri Hogue, Colorado School of Mines, Golden
Christopher Hoving, Michigan Department of Natural Resources, Lansing
Matthew McKinney, University of Montana, Missoula
Vicky Meretsky, Indiana University, Bloomington
Jensen Montambault, The Nature Conservancy, Charlottesville, Virginia
Virgil Moore, Idaho Fish and Game Commission, Boise
Lynne Sebastian, SRI Foundation, Rio Rancho, New Mexico

Although the reviewers listed above have provided many constructive comments and suggestions, they were not asked to endorse the conclusions or recommendations, nor did they see the final draft of the report before its release. The review of this report was overseen by **Susan Hanson,** Clark University, and **Bonnie McCay,** Rutgers University, who were responsible for making certain that an independent examination of this report was carried out in accordance with institutional procedures and that all review comments were carefully considered. Responsibility for the final content of this report rests entirely with the author committee and the institution.

The committee would like to especially thank Elsa Haubold and Karen Murphy from the U.S. Fish and Wildlife Service for their invaluable assistance in providing background documents requested by the committee throughout the report writing process. This report was also greatly enhanced by the efforts of those who participated in committee meetings and informal phone conversations. The committee would like to acknowledge Sergio Avila, Rob Campellone, Dan Decker, Carolyn Enquist, Lise Hanners, John Harja, Polly Hicks, Genevieve Johnson, Randy Johnson, Rodd Kelsey, Ken Mayer, Jerry McMahon, Marcia McNutt, Virgil Moore, Dan Odess, Brady Phillips, Karen Pletnikoff, Duane Pool, Jeff Raasch, Ron Regan, Sue Rodman, Glen Salmon, Ray Sauvajot, Lynn Scarlett, John Schmerfeld, Michelle Selmon, Allison Shipp, Carter Smith, Wayne Spencer, Buck Sutter, Gary Tabor, Stephanie Toothman, Larry Voyles, Gwen White, David Whitehurst, Michael Whitfield, Ulalia Woodside, and Mark Zacharias.

Contents

Summary

The U.S. tradition of conserving fish, wildlife, habitats, and cultural resources dates to the mid-19th century. States have long sought to manage fish and wildlife species within their borders, whereas many early federal conservation efforts focused on setting aside specific places as parks, sanctuaries, or reserves. Starting in the 1960s, several federal laws were passed to provide additional protection for individual species and particular natural resources. In recent decades, resource managers and scientists gained greater appreciation of broader ecosystem dynamics that extend beyond geographic or political boundaries, as well as the increasing stress on ecosystems due to human activities. The convergence of these trends points to the need for a conservation approach that focuses on the landscape more holistically and integrates across multiple jurisdictional boundaries, sectors, stakeholders, and conservation goals.

The landscape approach (see Box S.1) is particularly important where multiple jurisdictions are involved; where the threats to species, ecosystems, and cultural resources occur at large regional scales; and where biological and geomorphic processes span across ecosystems. A migratory elk population, for example, might spend the summer in high-elevation forests on public land, migrate through a mix of public and private lands and waterways, and overwinter on private agricultural lands at lower elevations, making it challenging to develop a comprehensive management plan for the population. Likewise, historic and archeological properties, as well as traditional practices and livelihoods, such as ranching, farming, or subsistence harvest, often span public, private, and tribal lands, requiring an integrated approach to management.

In 2009, in recognition of the benefits of conservation partnerships at the landscape scale, the U.S. Department of the Interior's Secretary Ken Salazar issued Secretarial Order No. 3289 to establish the Landscape Conservation Cooperatives (LCCs), a network of 22 individual, self-directed conservation areas covering all of the United States, including Pacific and Caribbean islands, as well as parts of Canada and Mexico (see Figure S.1). The intent of the Secretarial Order was to provide a collaborative framework that could deliver the scientific information needed for effective management, and catalyze conservation planning and actions across multiple jurisdictions through partnerships.

Based on this Secretarial Order, a program titled the *LCC Network* was established with the main objectives to facilitate collaboration across jurisdictional boundaries, develop shared conservation priorities and common science needs among partners, and create conservation strategies to be implemented by participating agencies or other partners. Each LCC has its own governance structure, coordinators, and steering committee, which develop strategic conservation priorities for the region. The 22 LCCs are coordinated by a small team at the U.S. Fish and Wildlife Service (FWS), with input from the LCC Council, an advisory group that helps shape the LCC Network's overall strategic vision and goals. The LCCs and the LCC Network were funded at roughly $14 million per year and received about $11 million for science support during the fiscal years 2014 and 2015.

The conservation community had a mixed response to the announcement and initial formation of the LCCs. Some regions eagerly embraced the concept, swiftly formed a steering committee, and identified joint conservation priorities. Others argued that a new federal program was unnecessary, pointing to existing efforts that appeared to address at least some of the same issues. Many expressed concerns that participating in LCCs would place more demands on already limited staff time and other resources. Nonetheless, all regions have proceeded to establish a conservation cooperative to engage stakeholders.

Five years after the Secretarial Order was issued, Congress directed the FWS to ask the National Academies of Sciences, Engineering, and Medicine to "evaluate: (1) the purpose, goals, and scientific merit of the program within the context of other similar programs; and (2) whether there have been measurable improvements in the health of fish, wildlife, and their habitats as a result of the program." In

BOX S.1
Definitions

Landscape scale: "The term 'landscape scale' can represent many different spatial scales depending on the resources being managed. . . . A 'landscape' is defined as a large area encompassing an interacting mosaic of ecosystems and human systems that is characterized by a set of intersecting management concerns. The landscape is not defined by the size of the area, but rather by the interacting elements that are meaningful to the management objectives. In addition, . . . the term 'landscape' encompasses watersheds and marine environments that match the above description" (for the purposes of this report, the committee adopts the definition of "landscape scale" from the President's Priority Agenda: Enhancing the Climate Resilience of America's Natural Resources [Council on Climate Preparedness and Resilience, 2014]).

Landscape approach: "The 'landscape approach' seek[s] to provide tools and concepts for allocating and managing land to achieve social, economic, and environmental objectives in areas where agriculture, mining, and other productive land uses compete with environmental and biodiversity goals" (Sayer et al., 2013).

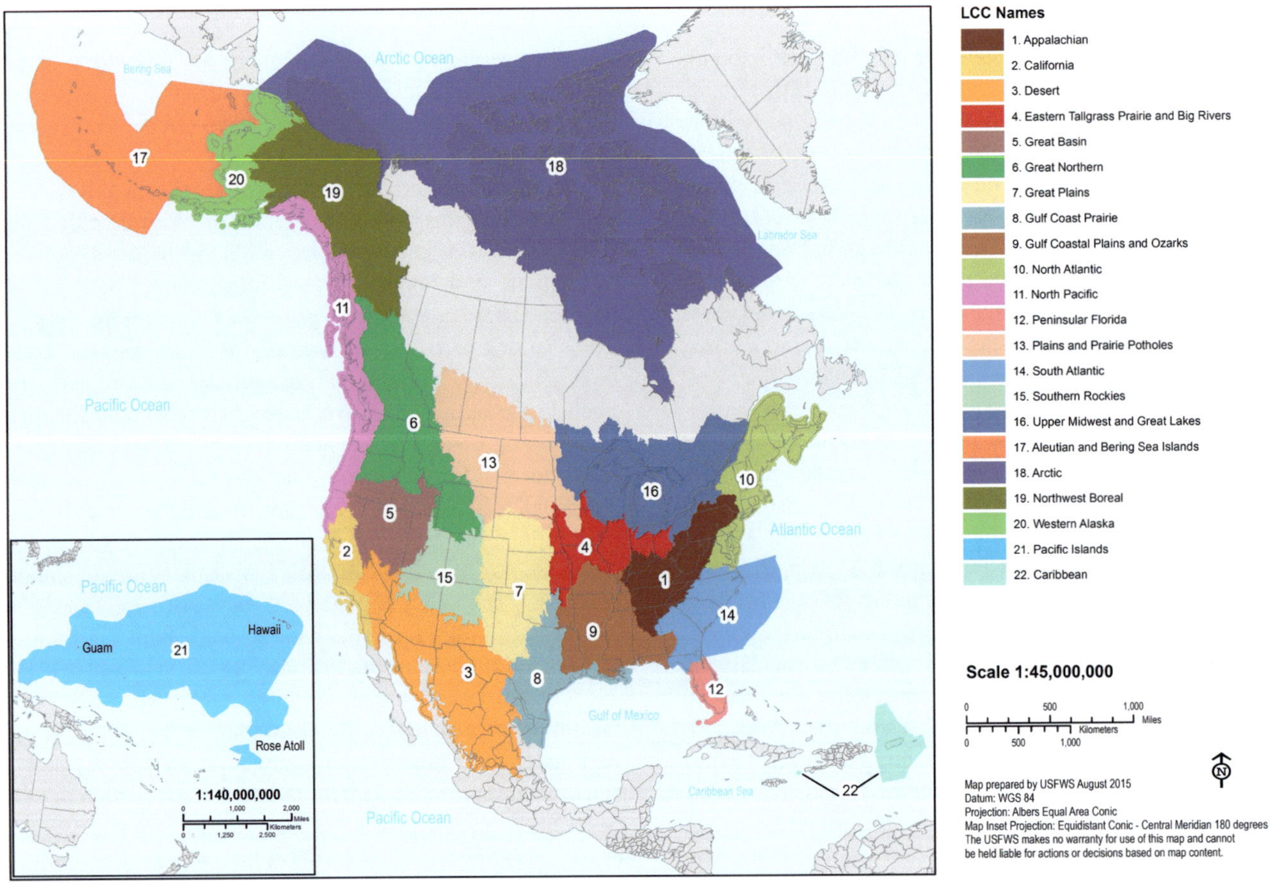

FIGURE S.1 Map of the Landscape Conservation Cooperatives. Each distinctly colored area on the map depicts the geographic coverage of each of the 22 individual, self-directed Landscape Conservation Cooperatives, which together cover all of the United States and parts of Canada, Mexico, and the Pacific Islands.
SOURCE: http://lccnetwork.org.

response, a statement of task was developed to both address this congressional request and provide an assessment of the evaluation process for LCCs (see Box S.2). The Academies established a study committee, which met numerous times, solicited extensive input from members of the stakeholder community, and authored this report.

NATIONAL NEED FOR A LANDSCAPE APPROACH TO CONSERVATION

Despite the strong rationale and broad support for landscape approaches in the conservation literature, many conservation activities in the United States continue to pro-

BOX S.2
Statement of Task

The National Academy of Sciences will convene an ad hoc committee to examine the Landscape Conservation Cooperatives (LCC) program. This committee will evaluate the purpose, goals, and scientific merits of the program within the context of similar programs, and whether the LCC program has resulted in measurable improvements in the health of fish, wildlife, and their habitats. This will include

1. An evaluation of the scientific merit of the LCC program and its goals.
2. A comparison of the stated purpose and goals of the LCC with other similar programs. How are these programs similar, and how do they differ? Is there substantial overlap in their mission and purpose? If so, is there rationale for and benefit from this overlap? Is there sufficient coordination with these related programs?
3. A comparison of the types of projects, activities, and collaborations supported by LCC and related programs.
   - Do the projects, activities, and collaborations supported by the LCC program overlap significantly with the traditional portfolio of other FWS programs (as the primary sponsoring agency)? Is there sufficient coordination and integration with these related programs? What benefit, if any, is gained by moving and/or consolidating this work within the LCC program? What effectiveness or efficiency is lost, if any, by housing this work within the LCC program? What changes can the FWS consider to address concerns?
   - Do the projects, activities, and collaborations supported by the LCC program overlap significantly with the portfolio of related programs in other agencies? Is there sufficient coordination with these related programs?
4. An examination of the evaluation process for the LCC program. What is the FWS's strategy to assess the effectiveness (output and outcomes) of the LCC program? What are reasonable short-, medium-, and long-term metrics for the effectiveness of the LCC program in achieving its stated purpose and goals?
5. An assessment of the impacts of the LCC program at various scales. What goals (and/or objectives) have been achieved? What improvements in managing and conserving habitat and fish and wildlife species might be reasonable to expect from the LCC program in the time frame it has existed? What longer-term impacts are likely to be realized?

ceed in a piecemeal fashion because of the way conservation programs and institutions have been established over time. Congressional and Executive Branch actions over the past century have created a complex tapestry of environmental and conservation policies intended to reverse the decline of the nation's natural and cultural resources. For example, the primary responsibility for species management falls to states; however, federal agencies are tasked with managing migratory birds, eagles, marine mammals, endangered species, and marine resources in the U.S. Exclusive Economic Zone. Furthermore, several federal statutes and regulations confer authority to federal agencies for managing and conserving the habitat on which species—managed by the states—depend. In addition, changes in species and landscapes have impacts on cultural resources, which fall under federal, state, local, or tribal authority. Often, critical habitat is privately owned, further impacting conservation strategies. For example, as wildfires in the western United States have become of increasing concern, differences in fire management policy across national parks, federally managed forests, tribal lands, state lands, and private lands can make it difficult to manage the risk of large fires, especially where the destruction of property and the potential for loss of human lives are at stake. These jurisdictional boundaries in management responsibilities raise some particular challenges with respect to the large number of species with ranges that cover federal, state, local, and private lands across multiple states. Given this complex web of management responsibility for natural and cultural resources, a collaborative approach to conservation is needed, especially in a time of sparse resources.

The LCC Network is part of the U.S. Department of the Interior's (DOI's) efforts to address this recognized national need. In 2014, the LCC Network published a strategic plan that articulates its goals, objectives, and representative tactics for the next 5 years (see Box S.3). The committee reviewed the high-level goals and objectives of this plan to assess whether the LCC Network has a strategy, structure, and function that can deliver on the promise of a landscape approach to conservation at the national scale. The committee concludes that the goals and associated objectives described in the LCC Network Strategic Plan are consistent with the research literature on landscape-scale conservation. The goals and objectives include most of the critical elements of a landscape approach, such as a unifying theme, a process for stakeholder engagement, adaptive management, and delivery of landscape-scale designs at the regional level with the aim to scale up to the network level. In general, the LCC Network Strategic Plan is consistent with the recognition that the most significant conservation challenges faced in the United States need to be confronted at a large spatial scale that transcends administrative and geopolitical boundaries and engages a diversity of stake-

BOX S.3
Goals and Objectives of the LCC Network Strategic Plan

"The LCC Network Strategic Plan identifies goals, objectives, and example tactics that support the Network's vision and mission. The goals identify common aims for individual LCCs and provide a way for them to align across the Network's geography" (LCC, 2014).

Four goals are identified in the LCC Network Strategic Plan:

Goal 1. Conservation Strategy: "An ecologically connected network of landscapes and seascapes adaptable to global change—such as climate change—with the ability to sustain ecological integrity and health to meet the needs of society at multiple scales."

Goal 2. Collaborative Conservation: "Facilitated alignment of partnership efforts within and amongst LCCs, including planning efforts and resources, that improves conservation outcomes across LCCs and the Network."

Goal 3. Science: "Natural and cultural resources are conserved at large landscape and seascape scales, guided by the collaborative application of science, experience, and cultural or traditional ecological knowledge and the generation of new conservation knowledge."

Goal 4. Communication: "Advance the knowledge of, support for, and engagement in landscape-scale conservation across the LCC Network."

holders across federal, state, tribal, local, and private land ownerships.

In examining the implementation of the LCCs to date, the committee concludes that individual LCCs were created to convene diverse stakeholders to work together across geopolitical boundaries to take on large-scale conservation challenges. LCC steering committee members include representatives from state agencies responsible for managing fish and wildlife species, from nongovernmental organizations, and from a range of federal agencies. These LCCs facilitate new opportunities to identify common conservation goals and priorities to leverage efforts of individual partners at much greater scale. Each LCC is engaging stakeholders within its region to identify and respond to the needs of local and regional partners. For example, several LCCs jointly produced the Gulf Coast Vulnerability Assessment, which is the first step in adapting to potential impacts of climate change. At the same time, the collection of LCCs is developing a national network to contribute to some of the federal agencies' national priorities.

The committee concludes that the nation needs to take a landscape approach to conservation and that the U.S. Department of the Interior is justified in addressing this need with the Landscape Conservation Cooperatives.

THE LANDSCAPE CONSERVATION COOPERATIVES IN THE CONTEXT OF OTHER SIMILAR PROGRAMS

Many programs within DOI as well as in other federal agencies also aim to coordinate conservation efforts and/or engage similar sets of stakeholders, leading to questions about the extent of overlap and coordination among these programs. The committee examined the goals and structure of many other federal programs and used the following five attributes to determine how similar they are to LCCs: extent of land coverage by the program, emphasis on research, emphasis on climate change as a driving issue, emphasis on natural and cultural resource conservation within the stated priorities, and extent to which the program's governance is concentrated in a single agency. These five selection criteria were chosen because the committee believes they sufficiently capture the essential attributes of the LCC Network.

Although many different programs and initiatives across federal agencies aim to undertake landscape-scale conservation, only a few programs overlap extensively with LCCs. Many other federal programs are more narrowly focused, with either a greater emphasis on research or a greater emphasis on conservation. Similarly, some of the federal programs span a much smaller geographic area, or are more narrowly focused on a single sector or a smaller set of target species. Generally, the committee found that there is good rationale for these overlaps and sufficient coordination with overlapping programs.

The committee concludes that the LCC Network is unique in that no other federal program is designed to address landscape conservation needs at a national scale, for all natural and cultural resources, in a way that bridges research and management efforts.

The programs with the most similarities to the LCCs are the National Oceanic and Atmospheric Administration's Regional Integrated Sciences and Assessments (RISA) program, FWS's Migratory Bird Joint Ventures program (Joint Ventures), FWS's National Fish Habitat Partnership, and the U.S. Geological Survey's Climate Science Centers (CSCs).

• The RISA program supports 11[1] regional research teams that help expand and build the nation's capacity to prepare for and adapt to climate variability and change. Broadly, RISAs focus more directly on adapting socioeconomic systems including urban water availability, public health, and community resilience to climate change. There is as much coordination between the LCCs and the RISAs as is reasonable to expect, given different heritages, structures, and focuses.

• The Joint Ventures were originally designed to develop conservation partnerships in support of the North American Waterfowl Management Plan. The conservation projects of the Joint Ventures are supported by contributions from many sources including but not limited to federal grants, state and corporation funds, and nongovernmental organization donations. The Joint Ventures share many similarities with the LCCs, in part because the LCCs were modeled after the Joint Ventures because of their successful track record in establishing partnerships and achieving conservation goals. The Joint Ventures consists of self-directed, regional partnerships that aim to develop shared priorities and goals. In general, they are more narrowly focused on migratory bird species and their habitats. The LCCs are thus much broader in scope, and each LCC has the flexibility to determine the target issues or species for conservation based on input from its steering committee members. Joint Ventures have membership on many LCC steering committees to help foster coordination.

• The National Fish Habitat Action Plan, which was developed in 2006 and updated in 2012, called for a network of regional Fish Habitat Partnerships that would focus on conservation of important aquatic habitats and species. The program is also modeled after the Joint Ventures and currently includes 19 regional, cross-jurisdictional Fish Habitat Partnerships comprising members from state and federal agencies, tribal governments, conservation groups, resource managers, academia, and other stakeholder groups. They operate regionally as self-directed partnerships and are collectively overseen by the National Fish Habitat Board. The Fish Habitat Partnerships have a narrower focus than the LCCs. The two programs have overlapping membership on their steering committees to ensure coordination and reduce redundancies.

• The CSCs were established through the same Secretarial Order as the LCC Network. Both programs have a geographic focus that spans the entire contiguous United States (although unlike the CSCs, the LCCs also include parts of Canada and Mexico), are intended to address the full range of conservation challenges, and are housed in DOI. The two programs were intended to be distinct but complementary with the CSCs having a greater emphasis on climate science delivery, while LCCs focus more on defining research needs for conservation and on science and tool delivery to support conservation outcomes. Both CSCs and LCCs award external grants, are guided by steering committees, include an emphasis on natural and cultural landscapes, and incorporate climate science to support decisions.

Of the programs reviewed, only the Joint Ventures and the CSCs appear to have some potential for redundancy that might need further consideration. In general, the LCCs have sought to enable coordination and reduce redundancies with these other related programs, for example, by having overlapping members on their steering committees. Each LCC has the flexibility to respond to its stakeholder conservation priorities, which makes it possible for the LCCs to avoid redundancies with the Joint Ventures and identify synergies. Despite the recognized need and unique niche for these multiple landscape-conservation partnerships, the number of such efforts can pose challenges to some partners whose active engagement is critical to achieving success.

Recommendation: DOI should review the landscape and habitat conservation efforts, especially the Joint Ventures and the LCCs, to identify opportunities for improved coordination between these efforts. Special consideration should be given to the limited capacity of state agency partners to participate in multiple efforts simultaneously.

The CSCs and LCCs were initiated to be complementary; strong coordination between the two programs was always intended and has been encouraged by having CSCs participate in LCC activities within their regions. In many cases the coordination between the two programs has been effective in helping bridge research investments and on-the-ground conservation needs. For example, many of the LCCs and CSCs run joint calls for proposals or at least consult during the design of a request for proposals. However, distinguishing the research funded by the two programs can be difficult, especially because both entities have evolved and aspire to fund scientific research and tool development in response to their respective stakeholder committees. Given this evolution, coordination becomes very important to ensure that the science funded by the two entities is complementary and not duplicative.

Recommendation: The LCC and CSC programs should be more clearly delineated. They should explicitly state how their research efforts differ and how they complement each other, identify and build on existing examples of coordination across the LCC Network, and make adjustments as appropriate. At the regional scale, LCC coordinators and CSC federal directors should coordinate their activities, including calls for proposals, as much as possible to avoid duplication of effort.

[1] Since this report entered review, the committee has learned that the Southeast Climate Consortium is no longer currently funded, though they will have an opportunity to apply for future funds again. Therefore the RISA program now currently supports 10 regional research teams, not 11. Because the committee learned of this after the report entered review, references to the Southeast Climate Consortium remain throughout this report.

EARLY ACCOMPLISHMENTS OF THE LCCs

The FWS developed the Science Investment and Accountability Schedule (SIAS) to evaluate individual LCCs. Each LCC is asked to conduct a self-evaluation using the SIAS instrument and submit it to the Network coordinator. This self-evaluation tool is currently the only formal LCC assessment that is conducted regularly. The committee reviewed the self-assessments of the individual LCCs and other materials provided by the LCC Network to evaluate what has been accomplished to date.

Despite the relative youth of the program, numerous objectives and milestones have been achieved by individual LCCs, especially related to developing partnerships and collaborative governance as well as identifying shared conservation goals. Each of the 22 self-directed LCCs appointed staff coordinators, science coordinators, and a steering committee; developed a governance structure; and convened a steering committee to develop a common set of goals. Most LCCs have identified shared conservation and research priorities for use by all partners, and a few LCCs are developing Landscape Conservation Designs. All LCCs and their steering committee members have initiated a critical first step in adapting to the impacts of climate change by developing vulnerability assessments. In fact, they all have developed vulnerability assessments for at least 33 percent of their geographic area and/or 33 percent of their priority resources.

In addition, progress has been made toward the LCC Network's goal to advance science; a considerable amount of scientific work has been funded and disseminated to resource managers. For example, 142 projects are funded under the "data acquisition and development" category, and 221 projects are funded under the "decision support" category. Although it was beyond the committee's scope to look at all of the research projects and activities undertaken by each LCC, the committee did hear from several stakeholders that a number of tools and research results have already improved resource management decisions and helped develop more cost-effective management options.

The committee concludes that the individual LCCs can point to many early accomplishments, and have made progress toward the LCC Network's high-level goals related to addressing conservation strategy, developing collaborative conservation, and advancing science for conservation.

EVALUATING PROGRESS FOR THE NATIONAL LCC NETWORK

The committee found it difficult to determine whether the LCC Network had resulted in measurable improvements in the health of fish, wildlife, habitats, and cultural resources. One significant limitation is the youth of the program. Most of the individual LCCs were initiated only 4 or 5 years ago, whereas conservation efforts typically take many years or decades to demonstrate measurable changes in the health of resources. Thus, it is too early to expect to see much improvement in the status of habitat, fish, and wildlife.

In addition, the LCC Network poses some unique challenges with regard to evaluating progress. Because the LCCs do not have the authority to deliver conservation actions, but instead rely on their partners to implement the on-the-ground management objectives, it becomes particularly challenging to measure conservation outcomes. In other words, while it is clear that improved coordination among these partners is needed to reach their shared conservation goals, it is not straightforward to measure the added value of the coordination. This, in turn, makes it challenging to measure and document how the LCCs contribute to improved management of fish, wildlife, habitats, and cultural resources. While acknowledging these challenges, it is important to attempt to articulate conservation measures for the purpose of evaluating and guiding LCCs' progress toward their stated goals.

Finally, it is important to recognize that the LCC Network must contend with a tension between demonstrating how individual LCCs meet the goals of their respective partners versus the goals of the federal government as required by legislation. The LCCs are designed to respond to their stakeholder needs; yet, as part of the federal government they are required to report on accomplishments to Congress. Thus, it is unclear at this point what evaluation and attribution process could link actions on the ground by partners of the LCCs to the planning effort of an individual LCC or the LCC Network as a whole to meet congressional requirements. In addition, it is not apparent to the committee how the strategic planning effort at the scale of the entire network of 22 LCCs (shown in Figure S.1) will result in actions on the ground.

Recommendation: The LCC Network should improve its evaluation process to better capture the contributions made by all partner agencies or groups toward common objectives. In particular, to demonstrate the effectiveness of the individual LCCs and the LCC Network, the evaluation process should measure how resources invested in any portion of the LCC Network further the goals of the LCC Network and their partners. The efforts invested in the LCCs and the LCC Network consist of (1) federal funding allocation via the FWS, (2) partners' in-kind contributions via staff time or technical expertise, and (3) funding from other state/federal agencies or private partners.

DEMONSTRATING BENEFITS TO LCC PARTNERS AND THE NATION

Ultimately, the LCC Network needs to demonstrate measurable benefits to its partners and the nation. This will require a framework that can connect needs and actions at a regional scale with conservation priorities identified at a network scale. The LCC Network has taken some steps in this direction by developing the LCC Network Strategic Plan and implementing the SIAS to evaluate individual LCCs. The committee has identified several ways that these efforts could be improved, particularly to address challenges in aggregating the achievements of individual LCCs to measure the progress toward network-wide goals.

An effective evaluation process must consider the programmatic goals, the reporting requirements, and the intended audience for the evaluation. Because the LCC Network is a federal program, its evaluation process needs to conform to guidelines set by the Office of Management and Budget (OMB) to demonstrate the performance of the program. It also needs to be able to evaluate the collaborative processes, which constitute a central programmatic goal of the LCCs and a relatively unique goal of a federal program.

The committee reviewed the goals and objectives in the LCC Network Strategic Plan and compared them to the benchmarks contained in the SIAS, and concludes that the SIAS provides a meaningful initial evaluation of the individual LCCs. Specifically, the performance metrics categories (process, outputs, and outcomes) in the SIAS conform with guidelines from OMB; the metrics align with some of the goals in the LCC Network Strategic Plan to some degree; and the metrics enable aggregation across the individual LCCs.

However, the SIAS tool falls short in the following ways: it is not based on the LCC Network Strategic Plan, and thus it does not assess progress toward the network goals and it does not measure how the LCCs and the LCC Network contribute to the goals of their partners. Most notably, the SIAS does not currently include a process to measure the outcomes that result from the collaborative process and its partners' on-the-ground conservation efforts. Although this FWS evaluation approach meets the agency's federal reporting requirements, the current metrics are not well suited for measuring the value added by the LCCs.

Recommendation: The FWS, in its next iteration of the SIAS, should (1) identify how and where the SIAS relates to elements of the LCC Network strategy; (2) identify the benchmarks associated with each Activity Area and continue that exercise by (a) classifying benchmarks as short, medium, or long term and (b) ensuring that benchmarks are adequately developed for and assigned to each SIAS Activity Area; and (3) begin the process of identifying, to the extent feasible and practicable, costs relative to returns on investment associated with achieving each benchmark.

The LCC Network Strategic Plan describes the network-wide goals and objectives, and therefore could provide a useful framework for metric-setting. However, the committee found that the plan does not always distinguish clearly between the conservation goals or objectives it is trying to achieve, and the processes and means by which it aims to achieve those conservation goals. There is some redundancy among plan "objectives" as well as some misplacement of objectives within main goals. As a result, the LCC Network Strategic Plan does not provide a clear hierarchy for goals and objectives. If revised, the plan could be used to develop an efficient set of metrics for evaluating LCC Network activities.

Developing metrics for the network as a whole may best be incorporated into an evaluation tool that is complementary to, but separate from, the SIAS. Because the SIAS is an FWS product that is reflective of the agency's own goals and objectives and is needed for federal reporting requirements, the LCC Network as a whole may consider an additional evaluation framework that can better capture goals, objectives, and measurements of the individual LCCs and allow aggregation toward network-wide progress evaluation.

Recommendation: Establishment of metrics at the individual and network-wide scales should become a high priority.

- Metrics should be developed to measure each LCC's unique goals, yet be consistent enough across LCCs to permit aggregation to a network scale.
- The criteria used to rate the performance of the LCC Network as a whole, and its components, should be closely related to the objectives that they are intended to evaluate and articulated clearly enough that any evaluator with access to the same information about the LCC Network could apply those criteria consistently.
- To more clearly demonstrate relevance to the stated purpose and goals of the LCC Network, as well as to better define the FWS role in support thereof, the SIAS Activity Areas and benchmarks should be written in a manner that clearly aligns with the LCC purpose and goals, as captured in the LCC Network Strategic Plan or its next iteration.

CONCLUDING REMARKS

The nation needs a landscape approach to conservation. Implementing landscape approaches in the United States is challenging because of the multitude of federal, state, local, and tribal jurisdictions, as well as numerous private landholders and stakeholders. The LCC Network initiated by DOI aims to address this national need. Many other programs are also striving to address regional conservation challenges. However, only the LCC Network is designed to address this need at a national scale for all natural and cultural resources, and to bridge from research to management.

As reported in the LCC self-assessments, individual LCCs have identified conservation priorities and undertaken many projects that will enable them to contribute to improved resource management. Evaluating progress at the network-wide scale has been more challenging because the program is still relatively new and because it needs better ways to measure the outcomes that result from the collaborative process and to link the conservation efforts undertaken by partners with network-wide goals and objectives. Improving their evaluation process will allow the LCCs and the LCC Network to demonstrate measurable benefits to their partners and the nation.

Assuming the LCC Network successfully implements its strategic plan, the LCCs will provide an important process by which stakeholders can engage at the landscape scale to set strategic conservation priorities that can span interest groups, disciplinary expertise, and sectoral approaches. This report provides guidance to further improve on the ability of the LCC Network to deliver on its vision of "landscapes capable of sustaining natural and cultural resources for current and future generations."[2]

[2] LCC vision statement: https://lccnetwork.org/about/about-lccs.

1

Introduction

The U.S. tradition of conserving fish, wildlife, habitats, and cultural resources dates to the mid-19th century. States have long sought to manage fish and wildlife species within their borders, whereas many early federal conservation efforts focused on setting aside specific places as parks, sanctuaries, or reserves. Starting in the 1960s, several federal laws were passed to provide additional protection for individual species and particular natural resources. Most states have similarly expressed a commitment to conservation in their constitutions and statutes, and over time, federal and state legislation has created a complex tapestry of legal frameworks for cultural and natural resource conservation. Conservation of habitats, species, ecosystem services, and cultural resources in the face of multiple stressors requires governance structures that can bridge the geographic and jurisdictional boundaries of the complex socio-ecological systems in which landscape-level conservation occurs (Jacobson and Robertson, 2012; Bodin et al., 2014).

Furthermore, in recent decades, conservation practitioners and scientists gained greater appreciation of broader ecosystem dynamics that extend beyond geographic or political boundaries, as well as the increasing stress on ecosystems due to human activities. Human activity has altered the surface of the Earth more substantially during the past half-century than at any time in history, and nearly half of Earth's land surface is used to grow crops and pasture animals (Foley et al., 2005; NRC, 2010). These changes in land use are threatening biological diversity, such that approximately 10–20 percent of species in well-known taxonomic groups are currently threatened with extinction (Pimm et al., 2014; IUCN Red List[1]).

With advances in landscape ecology over the past quarter-century, conservation planners, scientists, and practitioners began to stress the importance of conservation efforts at the scale of landscapes and seascapes (see Box 1.1 for a definition of landscapes and the landscape scale). These larger areas were thought to harbor relatively large numbers of species that are likely to maintain population viability and sustain ecological processes (e.g., migration) and natural disturbance regimes—often considered critical factors in conserving biodiversity (Schwartz, 1999; Soulé and Terborgh, 1999; Groves et al., 2002; Noss, 2002; Groves, 2003). Further, loss of biodiversity can have widespread consequences, including regional or global socioeconomic impacts or alterations to biophysical processes. By focusing conservation efforts at the level of whole ecosystems and landscape, practitioners can better attempt to conserve the vast majority of species in a particular ecosystem. Still, some species will not be well protected by a broad, ecosystem-level approach and will require species-level strategies (Noss, 1987; Hunter et al., 1988).

Because the loss of species and degradation of landscapes can reduce the availability of ecological goods and services on which humans depend, a balance between economic development and conservation of ecological integrity is needed (e.g., NRC, 2004). For example, pollinators are critically important to agricultural production, and the decline of honey bees in North America is a concern (NRC, 2007). The landscape approach to conservation aims to balance all human uses of a particular landscape (e.g., agriculture, ranching, energy development, hunting, recreation, etc.) with the need to conserve natural and cultural resources for future generations.

Successfully addressing the large-scale, interlinked problems associated with landscape degradation will necessitate a planning process that bridges different scientific disciplines and crosses sectors, as well as an understanding of complexity, uncertainty, and the local context of conservation work (Reyers et al., 2010; Curtin, 2015). The landscape approach to conservation aims to develop shared conservation priorities across jurisdictions and across many resources to create a single, collaborative conservation effort that can meet stakeholder needs.

Conservation scientists conclude that working at landscape scales is likely to be more effective for addressing current threats to biodiversity, such as widespread conversion

[1] See http://www.iucnredlist.org.

BOX 1.1
Definitions

Landscape scale: "The term 'landscape scale' can represent many different spatial scales depending on the resources being managed. For the purposes of this report, a 'landscape' is defined as a large area encompassing an interacting mosaic of ecosystems and human systems that is characterized by a set of intersecting management concerns. The landscape is not defined by the size of the area, but rather by the interacting elements that are meaningful to the management objectives. In addition, for the purposes of this report, the term 'landscape' encompasses watersheds and marine environments that match the above description" (for the purposes of this report, the committee adopts the definition of "landscape scale" from the President's Priority Agenda: Enhancing the Climate Resilience of America's Natural Resources [Council on Climate Preparedness and Resilience, 2014]).

Landscape approach: "The 'landscape approach' seek[s] to provide tools and concepts for allocating and managing land to achieve social, economic, and environmental objectives in areas where agriculture, mining, and other productive land uses compete with environmental and biodiversity goals" (Sayer et al., 2013).

of native landscapes for human use (e.g., agriculture, energy development, and urbanization), human population growth, and climate change (Franklin, 1993; Groves et al., 2002; Scherr and McNeely, 2008). This focus on landscapes and seascapes in conservation of natural and cultural resources is prevalent throughout the world today, including within federal agencies (FWS, 2013a), nongovernmental organizations (NGOs; e.g., Pressey and Bottrill, 2009; McKinney et al., 2010), and local land-use planning agencies (Marsh, 2010), and in the movements toward ecological networks such as the Natura 2000 network in Europe (Kati et al., 2015).

LANDSCAPE CONSERVATION COOPERATIVES

In response to the aforementioned large-scale challenges facing conservation planners and resource managers, the U.S. Department of the Interior (DOI) launched the Landscape Conservation Cooperatives (LCC) Network to enhance the landscape-level approach to conservation. Secretary Ken Salazar established the LCC Network on September 14, 2009, by DOI Secretarial Order No. 3289. Section 3(c) of the order reads as follows:[2]

> 3(c) <u>Landscape Conservation Cooperatives.</u> Given the broad impacts of climate change, management responses to such impacts must be coordinated on a landscape-level basis. For example, wildlife migration and related needs for new wildlife corridors, the spread of invasive species and wildfire risks, typically will extend beyond the borders of National Wildlife Refuges, Bureau of Land Management lands, or National Parks. Additionally, some bureau responsibilities (e.g., Fish and Wildlife Service migratory bird and threatened and endangered species responsibilities) extend nationally and globally. Because of the unprecedented scope of affected landscapes, Interior bureaus and agencies must work together, and with other federal, state, tribal and local governments, and private landowner partners, to develop landscape-level strategies for understanding and responding to climate change impacts. Interior bureaus and agencies, guided by the ***Energy and Climate Change Council***, will work to stimulate the development of a network of collaborative "Landscape Conservation Cooperatives." These cooperatives, which already have been formed in some regions, will work interactively with the relevant ***DOI Climate Science Center(s)*** and help coordinate adaptation efforts in the region.

Thus, the LCC Network was initiated to complement and add value to the many ongoing federal, state, tribal, and nongovernmental efforts to address a problem that this committee calls "the conservation challenge." This challenge is conserving species, habitats, ecosystem services, and cultural resources across jurisdictions, landscapes, and even national borders in the face of large-scale and long-term threats, including climate change.

IMPLEMENTATION OF THE SECRETARIAL ORDER: THE LCC NETWORK STRUCTURE AND FUNCTION

The intent of the Secretarial Order was to design a cooperative effort that can bridge jurisdictional boundaries across agencies within DOI as well as across other federal, state, and tribal agencies, and private lands. To implement this Secretarial Order, DOI charged the U.S. Fish and Wildlife Service (FWS) and other DOI bureaus with establishing the LCCs, and the U.S. Geological Survey (USGS) with establishing the Climate Science Centers (DOI, 2011).

Because of the FWS's experience with the Migratory Bird Joint Ventures program (Joint Ventures), the FWS used the Joint Ventures as a model to define the structure and function of the LCC Network. The Joint Ventures are collaborative partnerships organized to support conservation of migratory birds (for additional information on the Joint Ventures, see Chapters 5 and 6). The geographic boundaries for the individual LCCs were based on the National Geographic Framework for landscape-scale conservation that

[2] Material in bold italics reflects amendments made to this order on February 22, 2010.

was developed in partnership between FWS and USGS staff (Millard et al., 2012). Consequently, the network consists of 22 individual LCCs, which together cover all of the United States, including U.S. island territories and nations in the Pacific and Caribbean, as well as parts of Canada and Mexico (see Figure 1.1).

Once the geographic boundaries were determined, nine LCCs were established in 2010 and the remaining 13 were established in 2011 and 2012. Seven LCCs include Canadian membership and land, while three extend into Mexico. Each LCC has its own governance structure, including a steering committee (described further below).

The administration and staff for each LCC is currently supported financially by a federal or state agency. The FWS receives allocations within its budget for administrative and science support for the LCCs. For fiscal year 2014 (FY 2014), the FWS budget provided $14.4 million in funds for managing the LCCs and $10.8 million for science support. The FY 2014 budget also allocated 71 full-time employees for the LCCs. The FY 2015 allocations were roughly the same ($14 million for LCCs and $10 million for science support). In FY 2015, funding for individual LCCs was between $400,000 and $800,000 for each LCC including its science support. Some of the science support is pooled to fund multi-LCC efforts. For 17 LCCs, the FWS provides full-time equivalent funding for the LCC coordinator and the science coordinator. The other five LCCs also receive staff support from other federal agencies, namely the Bureau of Land Management, the Bureau of Reclamation, the National Park Service, and the U.S. Forest Service. State fish and wildlife agencies also directly contribute to the support of five LCCs.

Each LCC has a volunteer steering committee whose membership typically includes representatives from federal and state agencies, tribes, universities, the private sector, and NGOs. Each steering committee has a chair and a vice chair (or co-chairs), more than two-thirds of whom have been state members. The size of the steering committees varies, but most often ranges between 22 and 26 members; the proportion of federal membership in each steering committee varies from slightly under one-third to one-half of the total members. Steering committee members and staff formulate shared priorities and objectives, and most have developed

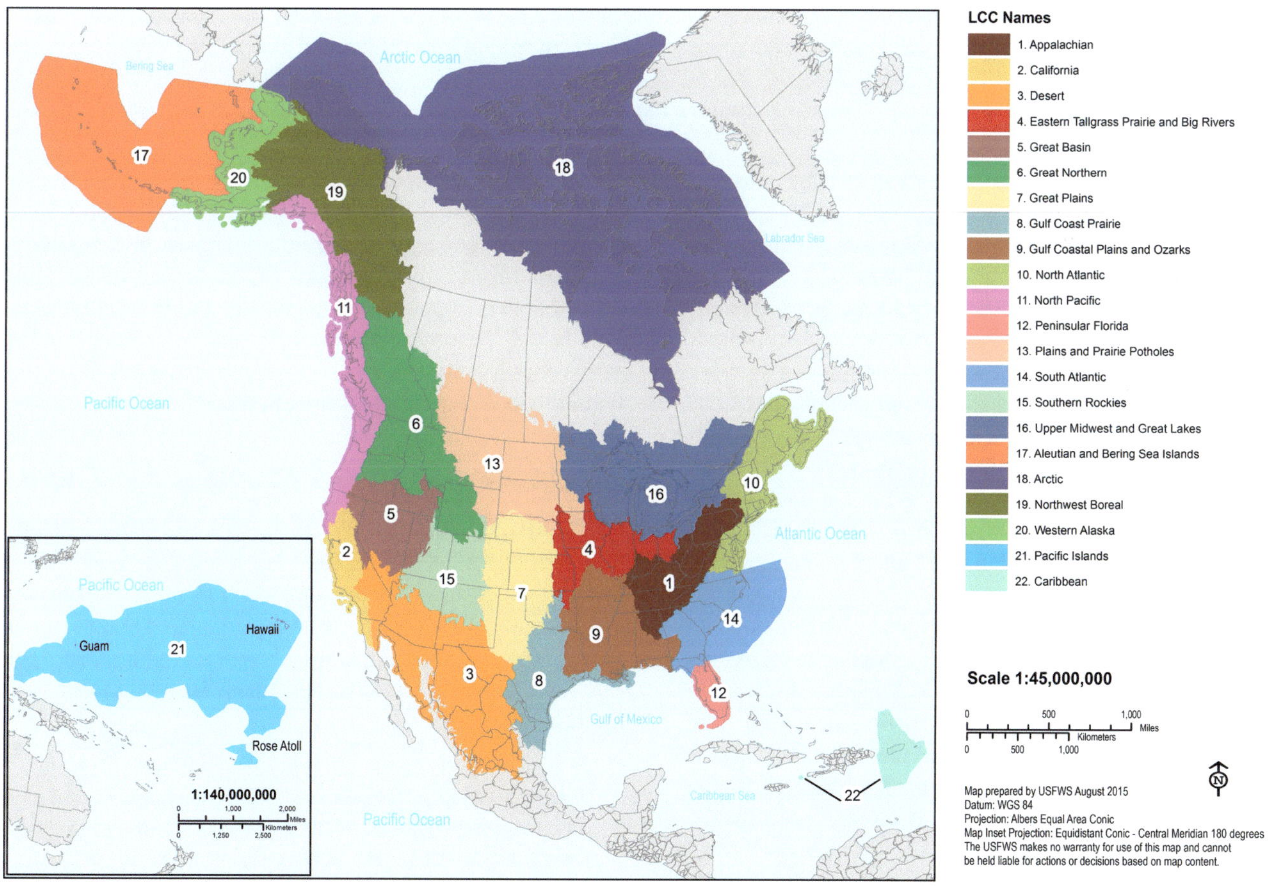

FIGURE 1.1 Map of the Landscape Conservation Cooperatives. Each distinctly colored area on the map depicts the geographic coverage of each of the 22 individual, self-directed Landscape Conservation Cooperatives, which together cover all of the United States and parts of Canada, Mexico, and the Pacific Islands.
SOURCE: http://lccnetwork.org.

BOX 1.2
Statement of Task

The National Academy of Sciences will convene an ad hoc committee to examine the Landscape Conservation Cooperatives (LCC) program. This committee will evaluate the purpose, goals, and scientific merits of the program within the context of similar programs, and whether the LCC program has resulted in measurable improvements in the health of fish, wildlife, and their habitats. This will include

1. An evaluation of the scientific merit of the LCC program and its goals.
2. A comparison of the stated purpose and goals of the LCC with other similar programs. How are these programs similar, and how do they differ? Is there substantial overlap in their mission and purpose? If so, is there rationale for and benefit from this overlap? Is there sufficient coordination with these related programs?
3. A comparison of the types of projects, activities, and collaborations supported by LCC and related programs.
   - Do the projects, activities, and collaborations supported by the LCC program overlap significantly with the traditional portfolio of other FWS programs (as the primary sponsoring agency)? Is there sufficient coordination and integration with these related programs? What benefit, if any, is gained by moving and/or consolidating this work within the LCC program? What effectiveness or efficiency is lost, if any, by housing this work within the LCC program? What changes can the FWS consider to address concerns?
   - Do the projects, activities, and collaborations supported by the LCC program overlap significantly with the portfolio of related programs in other agencies? Is there sufficient coordination with these related programs?
4. An examination of the evaluation process for the LCC program. What is the FWS's strategy to assess the effectiveness (output and outcomes) of the LCC program? What are reasonable short-, medium-, and long-term metrics for the effectiveness of the LCC program in achieving its stated purpose and goals?
5. An assessment of the impacts of the LCC program at various scales. What goals (and/or objectives) have been achieved? What improvements in managing and conserving habitat and fish and wildlife species might be reasonable to expect from the LCC program in the time frame it has existed? What longer-term impacts are likely to be realized?

strategic priorities in regard to conservation plans, studies, monitoring programs, infrastructure, tools, and expertise. LCCs aim to leverage resources, and partners typically contribute staff and financial resources or potentially contribute by developing and taking joint actions. Although the 22 self-directed LCCs need to be responsive to their own conservation issues, the program is intended to function as a network to respond to conservation challenges at broad multi-LCC scales too. Thus, while each LCC has unique characteristics and challenges specific to its geography, the individual LCCs also have the opportunity to scale up efforts and work with neighboring LCCs or other partnerships.

In 2010, the Udall Foundation's U.S. Institute for Environmental Conflict Resolution created an interagency LCC strategy team comprising federal, state, tribal, and NGO representatives to evaluate whether a higher-level policy body was needed to help move the LCC Network forward. There was consensus among the participants who then worked to develop a charter and select members for this new volunteer group, called the LCC Council, which met for the first time in 2014. To help provide a common vision for the LCCs, representatives from all 22 LCCs and network-level staff developed a network-wide strategic plan in 2014 building on the common themes across LCCs. The LCC Network Strategic Plan (LCC, 2014) lays out four main goal areas for the LCCs to address: conservation strategy, collaborative conservation, science, and communication. LCCs aim to provide decision support and conservation assessments, develop conservation priorities across jurisdictional boundaries, coordinate among agencies at different levels, and collaboratively develop monitoring needs. LCCs also provide funding for research in support of management and data integration. Finally, LCC partners can identify and contribute through actual conservation efforts and resource management actions within their authorities and jurisdictions.

ORIGINS OF THIS STUDY

This review of the LCC Network was congressionally mandated in 2014. Accompanying the amendments made by the U.S. House of Representatives to the U.S. Senate's amendments of H.R. 3547, the Consolidated Appropriations Act of 2014 was a report in the Congressional Record.[3] The report included the following language related to the appropriations for DOI:

> From within the funds provided for LCC activities, the [Fish and Wildlife] Service is directed to contract with the National Academy of Sciences to evaluate: (1) the purpose, goals, and scientific merit of the program within the context of other similar programs; and (2) whether there have been measurable improvements in the health of fish, wildlife, and their habitats as a result of the program.

[3] 160 Cong. Rec. H973 (2014).

Following receipt of this congressional directive, the FWS worked with the National Academies to develop a statement of task that would address the congressional request and respond to their needs for a comprehensive review of the LCC Network (see Box 1.2).

THE COMMITTEE'S APPROACH AND REPORT ROADMAP

The committee provides its comprehensive evaluation on the scientific underpinnings of the program, including relevant findings from the science of conservation and social science of resource management, in Chapter 2 (Statement of Task [SOT] Item 1). Given the focus of the congressional request on "improvements in the health of fish, wildlife, and their habitats," the committee's report focuses on LCCs' role in natural resource management, while still recognizing the need to include cultural resource management as part of the LCC portfolio as intended by the Secretarial Order. Chapter 3 reviews in greater detail the merit of the programmatic goals of the LCC Network Strategic Plan (SOT Item 1) and briefly discusses how the LCC Network structure and function can contribute to achieving such goals. To ensure effective programmatic evaluations in the future, the committee reviews the FWS's process for evaluation and provides guidance for setting effective metrics for the LCCs and the LCC Network in Chapter 4 (SOT Item 4). A review of related programs and the coordination among these programs is provided in Chapter 5 (SOT Items 2 and 3). Chapter 6 discusses some early achievements of the LCCs and likely long-term impacts and outcomes of landscape-scale conservation (SOT Item 5). The committee's main conclusions and recommendations are summarized at the ends of Chapters 2–5, while Chapter 6 includes conclusions in the main body as well as at the end. The committee provides two case studies in Appendixes A and B and provides guidance for developing Landscape Conservation Designs in Appendix C.

2

Scientific and Conservation Merits of Landscape-Scale Conservation and the Landscape Conservation Cooperatives

This chapter addresses Item 1 in the committee's statement of task: "an evaluation of the scientific merit of the Landscape Conservation Cooperatives (LCC) program and its goals." The committee interpreted this item as asking whether a national program focused on developing a landscape approach to conservation has scientific merit, and if so, whether the LCC program has made use of current, relevant science. Therefore, this chapter reviews the development and underpinnings of conservation science with a focus on landscape scales. It also discusses modern approaches to conservation with the aim of evaluating the degree to which the LCC program has incorporated this body of work.

HISTORY OF LANDSCAPE-SCALE CONSERVATION

The United States has a long history of conserving large areas even though many 19th- and early 20th-century wildlife conservation efforts in the United States were primarily focused at relatively local scales, and undertaken by state or federal authorities or individual landowners. The committee can point to a few examples of early, large-scale conservation efforts. The Migratory Bird Treaty Act of 1918 was intended to provide further protection for birds, in addition to earlier legislation that protected them from market hunting, and it was followed by the Migratory Bird Conservation Act of 1929.[1] That act, according to its preamble, was

> an act to more effectively meet the obligations of the United States under the Migratory Bird Treaty with Great Britain by lessening the dangers facing migratory game birds from drainage and other causes, by acquisition of areas of land and of water to furnish in perpetuity reservations for the adequate protection of such birds; and authorizing appropriations for the establishment of such areas, their maintenance and improvement, and for other purposes.

Note that this preamble expressly envisions establishing refuges in a landscape and the one threat it mentions—drainage—is a landscape-scale threat. Passage of the act led to the establishment of wildlife refuges on major flyways to protect migratory birds, an early implementation of landscape-scale conservation. The Taylor Grazing Act of 1934, which provided for management of grazing on public lands, also was a response to a landscape-scale threat, overgrazing on western rangelands.

Despite the above examples, many environmental threats at the time that conservation and systematic land management began continued to be perceived as local. However, many recent threats to natural resources occur at much larger spatial scales. Ecosystem degradation is occurring at an unprecedented rate due to resource extraction, urban expansion, air and water pollution, deforestation, agriculture expansion, and climatic changes. Whether it is wildfires over millions of acres, endangering human lives and property as well as economically and environmentally valuable resources, the cumulative impacts of thousands of dams disrupting natural flow regimes in large and small watersheds, the regional impacts of climate change on natural and cultural resources including people, or invasive species that alter the structure and function of entire ecosystems, it is clear that the most significant conservation challenges facing the United States today transcend administrative and geopolitical boundaries.

The scale of the conservation response does not yet match that of these threats. Scaling up conservation efforts and engaging a range of stakeholders across jurisdictions becomes necessary to muster a response proportional to the threats. Transboundary conservation also increases the capacity for finding solutions. For example, groups such as the Fire Learning Network facilitate structured learning across geopolitical boundaries primarily by establishing trust (Goldstein and Butler, 2012). This learning often brings benefits to conservation management at a local scale where conception, design, and building of local support may have been challenging without transboundary collaboration.

[1] Chapter 257, approved Feb. 18, 1929, 45 Stat. 1222.

Landscape-scale conservation is not an idea only of the past few years (Turner, 1989; Dunning et al., 1992; see Box 1.1 for definition of landscape). Scientists and resource managers have long understood that important threats to natural resources operate at a landscape scale, and so effective management may require a large-scale view of systems. For example, state agencies and nongovernmental organizations have developed a landscape-scale conservation effort for the greater sage-grouse (see Appendix A). However, the fact that landscapes span jurisdictional and institutional boundaries complicates efforts to implement conservation at the landscape scale. Here, we briefly describe the recent history of federal landscape-scale conservation approaches, both to show that the LCCs are not the first attempt to address this problem and to show that effective landscape-scale management requires careful attention to institutional design. No one entity is capable of carrying out landscape-scale management by itself. By its nature, landscape-scale conservation requires partnerships or at least coordination among the entities that share authority across a landscape. These partnerships can be both difficult to create and difficult to sustain over time.

Significant federal emphasis on a landscape approach to conservation dates back at least to the mid-1980s, but it became prominent in the 1990s. The first major landscape approach to conservation for the U.S. Fish and Wildlife Service (FWS) was the North American Waterfowl Management Plan, signed by the United States and Canada in 1986 (FWS and Environment Canada, 1986). Recognizing that cooperative harvest management efforts, which had been in place since the 1916 Convention for the Protection of Migratory Birds, were inadequate to protect the continent's waterfowl, the two nations agreed to undertake cooperative planning for habitat protection (for additional description of landscape-scale conservation at the FWS, see Chapter 5).

Enthusiasm for large-scale management efforts grew rapidly in the scientific community and among conservation practitioners (Agee and Johnson, 1989; Clark et al., 1991; Slocombe, 1993). Within a few years, spurred by a series of perceived crises under the Endangered Species Act (ESA) of 1973, regional ecosystem management had become a leading strategy in the federal conservation toolkit. The National Park Service's National Heritage Area program was initiated 30 years ago in recognition of the need to preserve cultural resources across the nation beyond the National Park Service's boundaries (see Chapter 6 for further discussion). Shortly after taking office, the Clinton Administration endorsed the concept (Gore, 1993). Regional plans were developed for protection of endangered species habitat in the forests of the Pacific Northwest, for restoration of the Florida Everglades, and for coastal sage scrub conservation in southern California (e.g., Frampton, 1996).

The landscape approach to conservation has been pioneered and embraced by nongovernmental organizations; community-based groups; universities; state, local, and tribal governments; and many other nonfederal actors. They have developed a variety of approaches to work across boundaries to achieve multiple objectives. The sage-grouse case study in Appendix A is a prime example. Similarly, The Nature Conservancy undertook its largest land acquisition to date with the Gray Ranch in New Mexico in 1990, and this effort evolved into a major multi-organizational landscape project known as the Malpai Borderlands Project (Sayre, 2005). By the late 1990s, The Nature Conservancy had launched scores of landscape conservation projects across the United States aiming to develop functional landscapes at multiple scales (Poiani et al., 2000). These efforts featured several common elements: all involved management beyond traditional political and institutional boundaries, emphasizing cooperation and coordination among participating entities; most focused on the production and use of needed scientific information and endorsed the concept of adaptive management, updating management strategies in light of learning over time.

CONSERVATION CHALLENGES AT THE LANDSCAPE SCALE

While the scale of conservation varies with the species and issues being considered, achieving landscape-scale conservation can be quite challenging. In part, this is because of the multiple uses that are demanded of many if not most landscapes and seascapes—to provide food, energy, water, and other goods and services for people while maintaining ecosystem functioning and biological diversity. The large area covered by most landscapes and seascapes, the complexity of the social and ecological communities within them, and the challenges of research designs provide a scientific basis for conservation (Eigenbrod et al., 2011; Baylis et al., 2015). These factors make landscape-scale conservation difficult (Lawler et al., 2014), and they require a broad scientific base that includes both natural and social sciences.

Climate change and its effects on sea level rise, on the survival of species and ecosystems, and on other aspects of human and natural systems provide challenges—but also incentives—for landscape-scale conservation efforts (e.g., NRC, 2010). The synergistic interaction of climate and land-use change (Ellis et al., 2010; Theobald, 2014) in the United States adds to the complexity of the challenge. Although anthropogenic features cover a relatively small portion of the western United States (13 percent), the human footprint disproportionately affects areas of high biodiversity such as valley floors and stream margins, further exacerbating ecosystem and wildlife management challenges (Leu et al., 2008). This may very well be an underestimate of the human footprint. For example, Forman (2000) argued that the ecological effects of roads extend outward for more than 100 meters, resulting in his estimate that about one-fifth of the U.S. land area is directly affected ecologically by public roads (see also NRC, 2005a). At the same time, land-use change, the further development of roads and other infrastructure on many landscapes, and the effects of climate change may also present opportunities for developing landscape-scale solutions.

The size and distribution of landscapes will vary by the mosaic of species, ecosystems, and human systems being considered, and by the set of relevant management concerns. For example, the greater sage-grouse (see case study on sage-grouse in Appendix A) currently occupies 56 percent of its historical range and was considered for listing under ESA, primarily because of the loss and fragmentation of its habitat (see Figure A.1), the sagebrush steppe landscape covering all or parts of 11 states. Wyoming contains 40 percent of the current greater sage-grouse population (Doherty et al., 2010) and the FWS (2010) described the greatest threats to the species as being poor grazing practices, wildfires, and agricultural conversion of sagebrush landscapes to grasslands. The threats are exacerbated by displacement from otherwise suitable habitat and functional habitat loss due to energy and other infrastructure developments. Other states that currently encompass the sage-grouse's range have far fewer greater sage-grouse, and the primary concern for the species in the western portion of its range is loss of habitat due to wildfire and the subsequent spread of invasive plants. The other end of the spectrum of landscape size and configuration is the example of the Cheat Mountain salamander, which is only found above an elevation of about 3,000 feet within five counties in the Allegheny Mountains of West Virginia (Pauley, 1993). Conservation of this species depends on forest management, and 46 of the 60 known populations occur entirely within the Monongahela National Forest, simplifying protection of the species and its habitat. So the size and configuration of landscapes from both a science and conservation perspective are greatly influenced by the conservation features that are the focus of research and management attention.

INSTITUTIONAL CHALLENGES OF LANDSCAPE-SCALE CONSERVATION

We have described the mismatch between jurisdictional boundaries and threats to natural resources. In addition, institutional mandates and responsibilities are divided and fragmented. Authority for conservation of natural resources has long been divided between the public and private sectors, between federal and state governments, and among individual agencies with distinct missions at both federal and state levels. Management efforts that cross these institutional boundaries inevitably face complexities and conflicts (although wisely assembled coalitions can be powerful forces for conservation).

A substantial challenge to conservation is fragmentation in land ownership. Land ownership in the United States carries with it control over the vegetation growing from the soil (except when these rights have been suspended through an alternative legal arrangement, such as easements for conservation or agriculture, or historical or cultural preservation, or when other rights, such as for mineral extraction, may affect private owners' ability to exercise domain over vegetation on their land). By contrast, states have authority over wildlife, except for federally protected species. Native American tribes may also have or share regulatory authority on tribal lands. As a consequence, even where states have conservation responsibility for fish and wildlife, they often do not have control over their habitats.

Under the Property Clause of the U.S. Constitution, the federal government has plenary authority over the lands it owns. Since the late 19th century, the United States has used that authority to dedicate lands to a variety of purposes, encompassing both conservation and resource extraction. Beginning in the early 20th century, the United States also enacted a series of conservation statutes dealing with wildlife. Today, the states retain primary regulatory authority over most wildlife in most places, but federal regulation is primary with respect to endangered species, migratory birds, and marine mammals. In some cases, the boundaries between federal and state authority are contested.

State management of wildlife populations is not without complexity as well. States have laws giving them authority over all or most species resident within the state; all states manage fishing and hunting of resident species within their borders, although hunting of migratory birds and management of other federal trust species is done collaboratively with the FWS as allowed by federal regulation. However, many species managed by states occur over relatively large ranges covering several states. For example, the greater sage-grouse occurs in 11 western states (see Appendix A). State laws differ, sometimes significantly, in how wildlife populations are managed. Although most state agencies own or control some wildlife habitat, the vast majority of habitat supporting state-protected wildlife is managed by the aforementioned patchwork of public and private ownership. Although these differences in state wildlife laws and management do not seem to pose any threat to the greater sage-grouse itself, they do need to be taken into account in any conservation effort that includes several states.

Within the state and federal governments, conservation authority is frequently further divided or distributed among agencies with different missions. Federal authority over endangered species is divided between the FWS, which is responsible for protecting terrestrial and freshwater species, and the National Marine Fisheries Service in the U.S. Department of Commerce, which is responsible for marine species and anadromous fishes (except in Maine, where the FWS has responsibility for Atlantic salmon in freshwater). Federal land management authority is divided to varying degrees among the U.S. Forest Service in the U.S. Department of Agriculture, and the Bureau of Land Management, the Bureau of Reclamation, the National Park Service, and the FWS in the U.S. Department of the Interior. The U.S. Department of Defense has extensive landholdings, and it engages in conservation when feasible given that its primary mission is to ensure readiness of the nation's military forces. (The U.S. Army Corps of Engineers, in the U.S. Department of Defense, does manage and operate considerable areas around its dams, reservoirs, and dikes without a primary mission of military readiness.)

Just as importantly, conservation efforts often come into conflict with competing goals for the use of land among both private landowners and other government agencies. For example, in the arid west, water management by state agencies and by the federal Bureau of Reclamation and the U.S. Army Corps of Engineers has been in significant tension with protection of endangered riparian and aquatic species for decades (e.g., NRC, 2012). State wildlife management agencies can find themselves in conflict with water and land management agencies. Even within individual agencies, there can be tensions between conservation efforts and other goals. The National Forests and Bureau of Land Management lands, for example, are managed for "multiple use," meaning that they are used for such activities as grazing, timber, recreation, and mineral extraction. Finding the right balance among those uses is often difficult and controversial, both within the management agency and among external constituents.

Conservation efforts at the scale of large landscapes and seascapes are frequently complicated by this range of stakeholders with divergent values and mandates, and who are deeply invested in the outcome. In addition to federal and state agencies, stakeholders include environmental organizations, industry associations, nearby property owners, communities economically dependent on recreation or resource extraction, and others, including faith-based organizations (Hicks et al., 2015). These stakeholders may have a legal right to participate in decision making. Even if the law does not provide them with explicit participatory rights, they may believe their interests deserve consideration.

There is growing evidence from theory and practice that bottom-up stakeholder-driven processes are more likely to achieve long-term conservation outcomes than top-down driven ones. This evidence applies a variety of perspectives, including those interested in the growing connections of nature conservation efforts and human well-being (Milner-Gulland et al., 2014), the U.S. experience in collaborative conservation efforts and adaptive management at the landscape scale (Scarlett, 2013), a global perspective on the goals of nature conservation (Mace, 2014), and that of the scientific researcher studying stakeholder involvement in conservation (Young et al., 2013). Two recent and important examples support this observation: establishing a migration corridor for pronghorn in the Greater Yellowstone ecosystem (Berger and Cain, 2014) and establishing a marine protected area network under the auspices of state legislation in California (Gleason et al., 2010).

LANDSCAPE APPROACH IN CONSERVATION

Although conservation biology has always had a strong focus on the maintenance of long-term, persistent populations of individual species, it has more recently recognized the importance of habitat conservation at larger scales, which requires a landscape-scale management approach (e.g., Sayer et al., 2013; Box 1.1). The landscape approach includes consideration of the complex jurisdictional environments, where ecological processes such as the migratory dynamics of a species, the hydrological flow regime of a river, or the spread of wildfire span many ecosystem types and geopolitical boundaries. An elk population or other migratory ungulates, for example, may spend the summer in high-elevation forest on public lands, migrate downward through a mix of public and private land, and then overwinter on private lands intensively managed for agriculture.

As scientists and managers address conservation at the landscape scale, we identify four overarching principles that need to be considered. First, there is a need to focus on both the biodiversity that occurs within these landscapes and the ecological processes and services that are derived from it. Second, species occur at different spatial scales and respond to landscapes in different manners, and these factors must be taken into account in the scientific methods used to assess the status of species. Third, science and management at landscape scales are complex and expensive, and require innovative and collaborative approaches to efficiently utilize the available resources. Fourth, which conservation features or targets are addressed (which species, communities, ecosystems, and/or processes), which threats are considered, what science is brought to bear (e.g., ecological, economic, social, and political), and how conservation and science challenges are conceptualized will have considerable influence on whether successful conservation outcomes are reached. Three key issues to consider in conceptualizing the landscape approach follow:

1. Conserving biological diversity and ecosystem processes (including services) at landscape scales. Doing so requires addressing multiple spatial, temporal, and jurisdictional scales. There is no single best methodology for integrating ecosystem services into landscape conservation plans, but there is an increasing number of guidance papers, frameworks, and tools for doing so that landscape practitioners can put to use (e.g., Kareiva et al., 2011; Bagstad et al., 2013).

2. Addressing landscape-scale variation for different priority issues. Different species perceive a landscape in different ways, depending on their life histories, associations with major vegetation types, and prey distributions, among other things (Fischer et al., 2005). Betts et al. (2014) proposed a species-centered approach using species-distribution models and land-cover data to measure landscapes. This variation in scale of species distribution and causes of the loss and fragmentation of habitat presents a challenge for the LCC process, because each LCC was established with fixed geographic boundaries (although there is a process for adjusting boundaries;[2] LCC Network, 2011). The LCC boundary delineations were a result of thoughtful discussion, but it was recognized that these boundaries were context dependent. Fish boundaries often do not apply to birds, or for management of large, migratory mammals. Cultural issues do not even appear to have been part of the mapping efforts. As such, boundaries will always be somewhat problematic. To address the need for flexibility depending on the particular conservation concern, the boundaries were meant to be permeable. Any program needs to be bounded and administered through some structure (regions, districts, states, basins, etc.), but flexibility in delineating and adjusting boundaries does not need to hinder the conservation activities of the program unduly.

3. Adaptive management at the landscape scale. Any program that intends to achieve conservation outcomes at the landscape scale will need to provide supporting scientific data, analyses, and tools (e.g., Curtin, 2015). Perhaps most important, these tools will need to be incorporated into an overall management approach, preferably following the principles of adaptive management. Although there are many definitions of adaptive management, two reports by the National Research Council (NRC, 2003, 2005b) discussed the concept in detail as it applies to the landscape-scale restoration of the Everglades of South Florida, distinguishing between active and passive adaptive management.

> Walters and Holling (1990) defined three general ways to structure adaptive management: (1) trial-and-error, (2) active adaptive management, and (3) passive adaptive management. According to these authors, the trial-and-error or evolutionary approach (also referred to as disjointed incrementalism by Lindblom, 1968) involves haphazard choices early in system management while later choices are made from the subset of choices yielding more desirable results. Active adaptive management strategies use the available data and key interrelationships to construct a range of alternative response models (scenarios) that are used to predict short-term and long-term responses based on small- to large-scale "experiments." The combined results of scenario development and experimentation are used by policymakers to choose among alternative management options to identify the best management strategies. Passive adaptive management is based on historical information that is used to construct a "best guess" conceptual model of the system. The management choices are based on the conceptual model with the assumption that this model is a reliable reflection of the way that the system will respond. Passive adaptive management is based on only one model of the system and monitors and adjusts, while in active adaptive management a variety of alternative hypotheses are proposed, examined experimentally, and the results applied to management decisions. (NRC, 2003)

Landscape Conservation Design (FWS, 2013a) is a landscape-scale approach to refuge system planning that is being adopted throughout the National Wildlife Refuge System and is a major objective in the LCC Network Strategic Plan (Objective 2 under Goal 1 on Conservation Strategy; LCC, 2014). The foundation of Landscape Conservation Design is the FWS Strategic Habitat Conservation Framework, which is essentially the FWS's version of an adaptive management framework. It is quite similar in nature to the adaptive management framework of the Conservation Measures Partnership (CMP, 2013), which is increasingly being adopted as a standard by conservation organizations. Although the Strategic Habitat Conservation Framework does not explicitly mention "active adaptive management" in its use of competing models and alternative hypotheses, it clearly is designed to reflect active, rather than passive, adaptive management.

SOCIO-ECOLOGICAL SYSTEMS IN LANDSCAPE CONSERVATION

Conservation science has long recognized the interdependence of people and nature (e.g., Marsh, 1864; Leopold, 1947) and the complexities of this relationship at different scales. In modern terms, society receives a variety of benefits (ecosystem services) provided by ecosystems, such as water, food, and fiber that we harvest; the regulation of climate, disease, disturbances, and the quantity and quality of water; and the aesthetic, spiritual, and recreational connections to the land and sea (MEA, 2005).

Much of the justification for landscape-scale conservation is based on an understanding of nonhuman species and their environments. However, there are equally valid reasons to invest in landscape conservation based on our understanding of how social systems work and interact with ecological systems, often referred to in combination as the *socio-ecological system* (see Figure 2.1). Socio-ecological systems are nested sets of social and ecological factors that interact to produce goods for society such as food, fiber, and drinking water (Berkes and Folke, 1998). Conservation planners are increasingly focusing on socio-ecological systems as they develop landscape-level conservation plans (e.g., Ban et al., 2013; Curtin, 2015).

The diversity of stakeholders who exist in any given socio-ecological system will have a considerable influence on the conservation goals and outcomes for the associated landscape. How well stakeholders are able to collaborate in working toward mutual goals will ultimately be a major factor in achieving conservation outcomes. As a result, "collaborative conservation" has become a central concept in large-landscape conservation (Margerum, 2008; Mitchell et

[2] In brief, the change must be proposed by a member of the steering committee of an LCC (the "initiating LCC"), the abutting LCC(s) also must agree to the change, and then the change has to be approved by the U.S. Department of the Interior Climate Change Working Group or the LCC Council once established and operational (it was not in 2011).

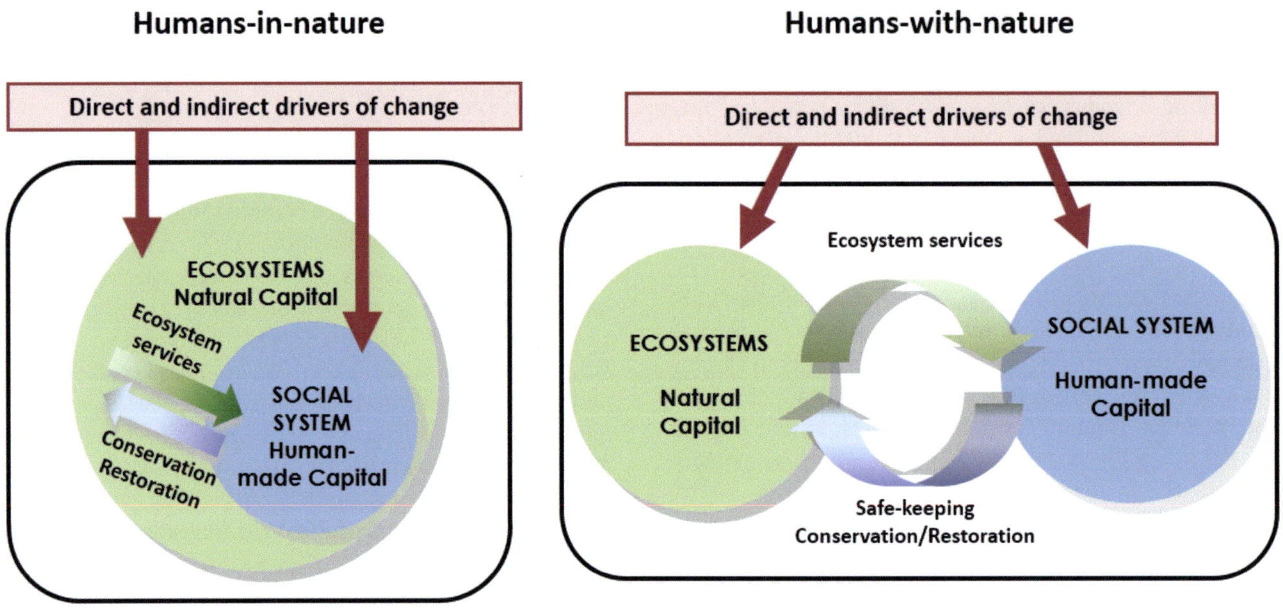

FIGURE 2.1 The socio-ecological system.
SOURCE: González et al., 2008.

al., 2015). In their ground-breaking policy report on large-landscape conservation, McKinney et al. (2010) noted 10 key elements of successful collaboration. Some of the most important elements include leadership, the representation of stakeholders, how the stakeholders make decisions and govern themselves, the strategies and outcomes that are deployed by collaboration, and the ability of a collaboration of individuals and institutions to learn and adapt.

Social scientists evaluating the implementation of State Wildlife Action Plans have developed a useful conceptual model of collaboration for landscape conservation. A key underlying assumption of collaboration in landscape conservation is that it requires learning among its participants in order to be successful (Lauber et al., 2011). Three types of learning appear to be important: social (relationships and dialogue among participants), conceptual (developing new ways to define problems and think about issues), and technical (developing new means to achieve objectives).

Cultures and livelihoods are just as sensitive to landscape processes and their jurisdictional controls as are natural resources, even though they can value aspects of the landscape differently. Nonetheless, cultural conservation also can benefit from a landscape perspective. Ranching, for example, often uses different landscape units spread across public and private lands to support summer and winter grazing. Subsistence and food security of indigenous communities usually involves harvest of fish and wildlife from landscape mosaics that cross multiple management jurisdictions. Farming often depends on groundwater recharge from complex landscape mosaics and influences nutrient delivery to downstream rivers and estuaries. Irrigation using surface water depends on a complex system of allocating water that considers native flows, stored water, priority water rights, and return flows, often extending throughout an entire river basin.

BOUNDARY ORGANIZATIONS

Some organizations are likely to be more effective than others in engaging in landscape conservation efforts. Those that may be more effective are ones that are able to bridge the knowledge-action boundary (Cash et al., 2003; Cook et al., 2013); that is, they understand the need for scientific information and data to inform conservation interventions, but they are also able to use such information in implementing conservation strategies and actions. In some cases, these organizations generate scientific information either directly or indirectly, but they also consume that information in advancing conservation actions that depend on the underlying science. Organizations that are able to bridge this knowledge-action boundary are referred to as boundary organizations and often employ or are able to facilitate dialogue across groups of scientists as well as natural resource managers and conservation practitioners who actually implement conservation. Nongovernmental organizations are often examples of boundary organizations because of their abilities to facilitate scientists, decision makers, and other stakeholder groups across landscapes (Cook et al., 2013). For example, the Ecosystem-Based Management Tools Network[3] "provides a wide range of training and outreach activities to connect practitioners with tools that incorporate

[3] See http://www.ebmtools.org.

natural and social science into decision making." The design and structure of LCCs suggest that they are intended to function as boundary organizations at large landscape scales by fostering collaboration among managers and stakeholders across multiple jurisdictions. It appears that in many cases, they may be able to fill this role, and may become more effective in fulfilling their missions if they are able to do so.

CHALLENGES OF LAUNCHING A LANDSCAPE APPROACH

Given the complex jurisdictional arrangements described above, it might not be surprising that the launch of the LCCs under the direction of the FWS encountered some challenges. First, this new program has been met with some criticism or skepticism. In its information-gathering process, during both committee meetings and informal phone conversations, the committee learned that the program was initially perceived by some stakeholders as a new federal mandate (as it was a Secretarial Order), as federal overreach, as a new program potentially decreasing funds for other, well-established FWS programs, and as yet another collaboration when there were already so many meetings to attend.

Because it was launched by a single agency in response to a Secretarial Order, its initial launch ran counter to the intent of the program and counter to the current social science research that stresses the importance of early involvement of relevant stakeholders to develop stakeholder engagement, trust, and buy-in (Cash et al., 2003). Thus, during the implementation, much of these initial perceptions needed to be overcome first before trust could be built. Furthermore, it appears that during the development of partnerships with individual LCCs, many of the same partners that the FWS typically engages were at the table. This could, in part, explain why the initial efforts by the LCCs appear to be primarily focused on natural resources, and why the methods by which the LCCs plan to address cultural resources are still not clear. As the LCCs evolve, benefits could be derived from engaging other state partners such as state parks, water agencies, and/or forestry agencies. This would serve to alleviate some of the burden of coordination among state agencies. The committee learned during its information-gathering process that during the process of engaging stakeholders and identifying their needs and priorities, a number of skeptical stakeholders began to recognize the potential benefits of this program.

CONCLUSIONS

Most of the major threats to conservation of natural and cultural resources in the United States today, including climate change impacts, occur at large spatial scales across vast regions, often as a result of historical or current land-use practices. To respond to these challenges, it is necessary to understand the underlying science of large landscapes and seascapes, and for different stakeholders and constituencies to work together across federal, state, local, and private jurisdictions. No single government agency, corporation, or nongovernmental organization is capable of doing this alone. The sage-grouse case study in Appendix A, based on a broad partnership initiated at the state level, illustrates the benefits from a landscape approach that helped avert the need for listing the species under ESA.

LCCs were created to convene different stakeholders to work together across geopolitical boundaries to take on these large-scale conservation challenges, including providing and further developing the underlying science capacity needed to address these problems. The committee concludes that the LCCs are appropriately based on modern conservation science and that, in concept, their design and implementation (with the exception of their initial launch) to date reflect an appropriate scientific response to conservation challenges at large scales.

The committee also concludes that the LCC boundaries appear to be largely fixed, in contrast to the definition of a landscape approach, which depends on the resources being managed. However, there is great emphasis in the LCC Network Strategic Plan to identify and address priorities that span multiple LCCs. One way the LCC Network has dealt with this matter is through initiating joint activities involving from two to seven LCCs (see Appendix B and further discussion in Chapter 6). In addition, the LCC Network has a process that allows for LCC boundaries to change under some circumstances (LCC Network, 2011; see footnote, p. 19). The committee applauds this flexibility, and encourages the LCC Network to explore the degree to which short- or long-term boundary changes can be helpful in achieving the program's mission.

3

Evaluating the Landscape Conservation Cooperatives Network Strategic Plan

The committee was asked to assess the scientific merit of the Landscape Conservation Cooperatives (LCC) Network and its goals. In this chapter, the committee discusses its approach to program evaluation in general, followed by an evaluation of the goals and objectives outlined in the LCC Network Strategic Plan (LCC, 2014; Appendix F). The committee evaluates this plan's goals and objectives based on value-focused thinking (Keeney, 1992). Furthermore, the chapter briefly discusses the structure and function of the LCC Network. The committee was also tasked to review the network's evaluation process, which is addressed in Chapter 4.

EVALUATING THE GOALS AND OBJECTIVES OF THE LANDSCAPE CONSERVATION COOPERATIVES

Meaningful evaluation requires clearly articulated and measurable objectives that reflect the mission of the program being evaluated, and that are sensitive to the context for which the program was initiated. Three kinds of objectives are relevant to evaluation: ends, means, and process objectives (Keeney, 1992; see Box 3.1).

The description of the LCC Network[1] can be parsed to show ends, means, and process objectives:

> With the signing of Secretarial Order No. 3289 [see Appendix E], the Department of the Interior launched the Landscape Conservation Cooperatives (LCCs) to better integrate science and management to address climate change and other landscape scale issues. By building a network that is holistic, collaborative, adaptive, and grounded in science, LCCs are working to ensure the sustainability of our economy, land, water, wildlife, and cultural resources.
>
> The 22 LCCs collectively form a network of resource managers and scientists who share a common need for scientific information and interest in conservation. Each LCC brings together federal, state, and local governments along with Tribes and First Nations, non-governmental organizations, universities, and interested public and private organizations. Our partners work collaboratively to identify best practices, connect efforts, identify science gaps, and avoid duplication through conservation planning and design.
>
> **Vision**
>
> Landscapes capable of sustaining natural and cultural resources for current and future generations.
>
> **Mission**
>
> A network of cooperatives depends on LCCs to:
>
> - Develop and provide integrated science-based information about the implications of climate change and other stressors for the sustainability of natural and cultural resources;
> - Develop shared, landscape-level, conservation objectives and inform conservation strategies that are based on a shared scientific understanding about the landscape, including the implications of current and future environmental stressors;
> - Facilitate the exchange of applied science in the implementation of conservation strategies and products developed by the Cooperative or their partners;
> - Monitor and evaluate the effectiveness of LCC conservation strategies in meeting shared objectives;
> - Develop appropriate linkages that connect LCCs to ensure an effective network.

Distinguishing properly between means and ends is important because there may be many ways of successfully achieving a desired end, and it would be unfortunate to preclude achieving such end objectives by too closely prescribing the means by which an ends objective should be pursued. In the LCC context, this is particularly important

[1] See http://lccnetwork.org/about.

BOX 3.1
Value-Focused Thinking

Value-focused thinking (Keeney, 1992) describes three types of objectives:

- **Ends** objectives express the underlying desirable qualities that the program is intended to enhance or create.
- **Means** objectives express ways of bringing those desirable qualities about through action.
- **Process** objectives express desirable qualities of the ways that decisions are made in the course of implementing a program.

Value-focused thinking encourages distinguishing as clearly as possible between objectives that are desirable in themselves, the ends, and objectives that are desirable because of their anticipated effect on accomplishment of desired ends. Both means objectives and process objectives are valuable in the latter sense. However, it is sometimes, and perhaps often, the case that process objectives may express some qualities that are desirable for their own sake, as well as being desirable for their anticipated effect on ends. For example, people value how decisions are made highly enough that a decision process that is perceived to be fair and responsive may even outweigh a substantively poor outcome in terms of overall satisfaction (e.g., Lind and Tyler, 1988). Similarly, stakeholder engagement in the development of conservation priorities can be viewed as a process, but in the context of the LCC program it is desirable for its own sake and can be viewed as an ends objective. These examples illustrate that developing and distinguishing objectives using this value-focused framework explicitly acknowledges the stakeholders' biases and value judgments regarding whether an objective is a means, ends, or process objective.

The value-focused thinking distinctions among ends, means, and process objectives are not entirely congruent with the way the LCC planning documents use the terms "goals" and "objectives," nor with the way that the Office of Management and Budget (OMB) uses the terms "outcomes," "outputs," and "process." OMB's term "outputs" roughly corresponds to "means"; OMB's term "outcomes" approximates the term "ends." Further discussion on the different use of these terms is provided in the text below.

because of the considerable autonomy of the regional LCCs in deciding how to operate within the overall LCC mission. It is important that the LCCs view certain process objectives as ends in themselves, as well as means, because of the emphasis that the overall LCC mission places on collaboration and partnerships in both the structure and the function of LCCs.

A hierarchical organization of ends objectives (and sub-objectives) is useful (1) because it helps to eliminate redundancy and identify gaps in program goals, and (2) because it leads directly to an efficient set of metrics that can be used for program evaluation (see discussion on metrics in Chapter 4). The LCC Network description quoted above includes only a few words about ends. To use these high-level, broad ends objectives to guide program evaluation requires elaboration of the aspects of each of the high-level objectives into sub-objectives. The LCC language offers some clues for sub-objectives, mentioning "economy, land, water, wildlife, and cultural resources" as being desirable aspects of sustainable landscapes. Each of those sub-objectives might be divided further, answering questions such as, What aspects of the economy are important in this context? Other possible sub-objectives are not so clear; for example, what are the elements of "sustainability" that are important? A fully expanded objectives hierarchy would identify the aspects of each higher-level objective or sub-objective, culminating in very specific criteria (i.e., metrics and measures) for evaluating the performance of programs to enhance achievement of the objectives in the hierarchy (where the activities of those programs are the means).

The LCC Network Strategic Plan (LCC, 2014; hereafter referred to as the strategic plan) is organized around four primary goals: conservation strategy, collaborative conservation, science, and communication. The strategic plan usage of "goals" versus "objectives" is quite common, but different from the value-focused usage of these terms described above. The strategic plan uses "goal" to describe high-level objectives, some of which are ends and some of which are means in the value-focused thinking. For each enumerated goal, the strategic plan identifies multiple objectives. The strategic plan uses "objective" to mean specific actions or targets needed to accomplish a goal; this usage is close to the *means* objectives of value-focused thinking.

The critique of the strategic plan that follows draws on the principles of value-focused thinking to analyze redundancies and potential inconsistencies in the strategic plan's formulation of goals and objectives as a first step toward developing an efficient set of evaluation metrics. The process of setting metrics is discussed in detail in Chapter 4. For each of the LCC Network's four goals (conservation strategy, collaborative conservation, science, and communication), the committee first analyzes the goal itself with respect to its status as an end in itself or a means to a desired end. Then, we present the objectives associated with that goal in a table, followed by analysis of the status of those objectives with respect to ends, means, and process.

Conservation Strategy

Devising a conservation strategy can be considered the overarching goal (*end*) for the LCC Network to pursue because the purpose of the entire program is to improve conservation effectiveness. The remaining three goals are really "means" to achieving this end, although to some degree they also can be viewed as ends in themselves. The full goal, as expressed in the plan, is to attain "[a]n ecologically connected network of landscapes and seascapes adaptable to global change—such as climate change—with the ability to sustain ecological integrity and health to meet the needs of society at multiple scales." We note that this is not a strictly scientific goal—"ecological integrity" and especially "ecological health" are not definable without reference to human value judgments (De Leo and Levin, 1997). Indeed, meeting human needs, which are associated closely with values, is part of the overarching purpose of the LCCs. This is an appropriate task of landscape-scale conservation, and it indicates the importance of obtaining broad agreement on which human needs and values will be considered when choosing actions and how progress toward meeting those needs will be measured. The objectives included in the strategic plan address the need to consider and identify a range of stakeholder priorities (and are *means* to the desired *end*; see Table 3.1).

All five objectives describe actions to be taken to accomplish the goal of landscape-level conservation and they are all means to this desired end.

Objective 1-1. As part of this objective, the strategic plan (see also Appendix F) specifies two example tactics: "establish conservation objectives at LCC and other applicable scales" and "roll-up LCC objectives to identify Network-scale objectives." While these tactics do not add much information, they do suggest a sequence of activities: first develop objectives at the LCC scale and then see if network-scale objectives emerge or can be extracted. Objective 1-1 is critically important. As discussed in Chapter 2, how well this particular means objective is achieved plays a critical role in realizing conservation outcomes.

Objective 1-2. The strategic plan's example tactics to promote resiliency and adaptation include working with various partners and indigenous peoples to identify "flagship regions" and priority areas for conservation; to provide tools and guidance for implementing and supporting conservation designs; and to acknowledge the extensive knowledge and practices of indigenous people. Objective 1-2 describes ways of communicating and supporting the results of Objective 1-1. As discussed in Chapter 6 and Appendix C, this objective is important to achieving the broader conservation goal of the program. It is not clear what the role of stakeholders is in disseminating this information, nor is it clear how the extensive knowledge and practices of indigenous peoples will be incorporated into the conservation designs, rather than simply being acknowledged.

Objective 1-3. Objective 1-3 is the program's means to scale up Landscape Conservation Designs to achieve conservation goals at the network scale. This task is an important rationale for creating the LCCs: to improve capacity for addressing broad-scale and cross-boundary problems that are unlikely to be addressed effectively through existing institutions and mechanisms. Further emphasis on Objective 1-3 will also help clarify the connection between each LCC's goals and regional- to network-scale goals (see discussion below). This consideration deserves more attention than has been provided to date, as discussed below.

TABLE 3.1 Brief Description of LCC Network Objectives for the First Strategic Goal, "Conservation Strategy"

LCC Network Goal- Objective #	LCC Network Objective Description
1-1	Identify shared conservation objectives, challenges, and opportunities to inform landscape conservation at continental, LCC, island, and regional scales.
1-2	Develop then deliver (through partners) regional landscape conservation goals and designs[a] that support resiliency and adaptation to both global change and regional landscape challenges, while ensuring the inclusion of all partners and stakeholders necessary for successful conservation.
1-3	Integrate regional or other scale-specific conservation designs to align and focus conservation action at the network scale, within available authorities.
1-4	Identify and obtain the resources required at the LCC and network scales to inform, develop, and support implementation of the conservation designs and other conservation actions.
1-5	Monitor the effectiveness of conservation design(s) and design application in terms of achieving stated outcomes, then revise as appropriate.

[a]Landscape Conservation Design is an iterative, collaborative, and holistic process that provides information, analytical tools, spatially explicit data, and best management practices to develop shared conservation strategies and to achieve jointly held conservation goals among partners.

Objective 1-4. This objective is critically important because work cannot be done without resources, which are means to the desired ends. However, given the current budget environment, resources dedicated to LCC activities through specific appropriation from Congress can be perceived as reducing funds available to other goals deemed important by partner agencies. Therefore, objectives under this goal might benefit from a careful review and adjustments in order to achieve broad stakeholder buy-in on Objective 1-1 prior to proceeding with Objective1-4 (see also discussion on program life cycle, below).

Objective 1-5. Objective 1-5 is an important *process* objective, as it expresses a commitment to adaptive management. A critically important part of that is self-evaluation, an assessment of whether and to what degree the program has been effective in (a) developing the conservation design and (b) yielding desired outcomes from the conservation design. This process objective is one of the key factors to successful landscape-scale conservation as outlined in Chapter 2 and discussed in greater detail in Chapter 6. Note that the type of adaptive management described in this objective is "passive," rather than "active," adaptive management, where contrasting management strategies are employed in a quasi-experimental manner in order to more efficiently learn how management actions create resulting ecological conditions.

Collaborative Conservation

The full statement of the collaborative conservation goal is "facilitated alignment of partnership efforts within and amongst LCCs, including planning efforts and resources that improve conservation outcomes across LCCs and the Network." This goal is intended to implement the "collaborative" element in the program's title. It is a means to achieving the conservation ends of the LCC program, although achieving collaboration also is an end in itself, given the LCC emphasis on partnership. It is a *process* objective in the value-focused thinking framework. The goal has six objectives (that are themselves *means*). As mentioned in Chapter 2, "collaborative conservation" is a central element of large-landscape conservation, and achieving this process objective and developing high-quality collaboration will be critical to the success of the program. Under this goal, the strategic plan lists six objectives (see Table 3.2).

Objective 2-1. It is important to develop and maintain a functioning network with the individual LCCs. Example tactics from the strategic plan include identifying barriers to collaboration and seeking to break them down, and continued pursuit of partnership opportunities. This objective reflects a continuing effort to develop the collaborative network that was announced by Secretary of the Interior Ken Salazar in 2010.

Objective 2-2. This objective includes a sample tactic to identify opportunities for cooperation across multiple LCCs, an approach that would help address conservation problems that frequently span multiple LCCs. It is unclear to the committee what process is in place to facilitate the identification of such opportunities and how the individual LCCs or the network as a whole would measure progress toward this objective.

Objective 2-3. This objective lists an example tactic to develop common definitions and performance metrics for key qualitative and quantitative outcomes that demonstrate value. Measuring effectiveness of the network is essential. However, as discussed in greater detail in Chapter 4, the LCC Network has not yet developed an effective process to monitor or evaluate the value and effectiveness of the LCC Network. As a result, it is difficult to scale from the effectiveness of individual LCCs to the network-level value and effectiveness. Achieving this process objective will be critical to sustaining support for this network.

Objective 2-4. This means objective appears to only minimally support the collaborative conservation goal, and does not appear to be a measurable objective.

Objective 2-5. The sample tactic indicates that this means objective focuses on seeking opportunities to leverage LCC funding. It is unclear how this would contribute to the goal of collaborative conservation.

Objective 2-6. In contrast, Objective 6 has no example tactic. It is related to the first two objectives under Goal 1, but

TABLE 3.2 Brief Description of LCC Network Objectives for the Second Goal, "Collaborative Conservation"

LCC Network Goal- Objective #	LCC Network Objective Description
2-1	Create a high-functioning organizational culture for LCCs and the LCC Network.
2-2	Identify and explore opportunities for collaborative actions within the LCC Network.
2-3	Demonstrate, monitor, and evaluate the value and effectiveness of the LCC Network.
2-4	Identify new and existing training and networking opportunities for the LCC Network.
2-5	Leverage conservation planning to be opportunistic in taking advantage of current and new funding sources for conservation.
2-6	Create a network-level system for prioritizing operational needs at network and regional levels, as appropriate.

focuses on operational needs rather than conservation goals. Objective 1-3 and this objective help with scaling up efforts and outcomes beyond the individual LCCs and are discussed in greater detail in Chapter 4.

Science

The full statement of the goal regarding science and the LCC Network is the following: "Natural and cultural resources are conserved at large landscape and seascape scales, guided by the collaborative application of science, experience, and cultural or traditional ecological knowledge and the generation of new conservation knowledge." The first phrase expresses an *end* and the remaining phrases express *means* of achieving that end. This goal has three *means* objectives (see Table 3.3).

Objective 3-1. Example tactics specified in the strategic plan include completing, disseminating, and implementing the LCC Science Plan; identifying and developing spatial, biological, and cultural data and evaluation tools across the network; and supporting assessment tools for climate change and climate adaptation planning for important cultural and subsistence resources. The goal identifies science as important, and the objective, and especially the example tactics, focuses on cultural and subsistence resources.

Objective 3-2. The example tactic to promote collaborative science is to include resource managers, cultural practitioners, and indigenous peoples in framing resource conservation issues and management objectives. This objective reflects an important recognition that environmental problems are defined relative to particular stakeholders; they are not abstract and self-contained issues. The objective and example tactic come closer than those mentioned earlier to recognizing the importance of incorporating the various stakeholders and areas of knowledge and expertise in all stages of the process.

Objective 3-3. The example tactic focuses on cooperating, sharing, cooperative synthesis of data, and communication. The objective is similar to previous ones that focus on evaluating effectiveness of the program. However, this objective includes both means and ends components because it aims to evaluate the effectiveness (means) and improve the effectiveness (ends) of LCC science. Perhaps most crucially, implementing this objective requires performance metrics to assess "the value of the LCC science." These metrics ideally align with the *ends* objectives, but that is complicated when *means* and *ends* are intermingled, as is the case here.

Communications

The full statement of the communication goal is to: "advance the knowledge of, support for, and engagement in landscape-scale conservation across the LCC Network." In other words, the goal appears to be to build a constituency for the LCC Network across all of its parts. It appears to be a *means* to the overall *end* of improving landscape-scale conservation. The goal has five *means* objectives (see Table 3.4).

Objective 4-1. The example tactic given in the strategic plan includes use of the LCC Network website and other tools as platforms for sharing news and the value of the program. This objective is a restatement of Goal 4, and in this restatement, the broader user community outside the LCC Network is included, as well as the network and the individual LCCs. Clearly, implementing this goal is important for at least two reasons. First, no matter how good the products of the LCC Network might be, their usefulness will be diminished to the extent that they are not widely known. Second, without outside knowledge of the program and its usefulness, continued support of the program will be difficult. The LCC Network has a challenge in promoting its successes and usefulness without detracting from the activities of other programs that it facilitates, partners with, and depends on.

Objective 4-2. The example tactic is to identify new, strategic target audiences whose interests might intersect with conservation interests. As discussed in Chapter 2, landscape-scale conservation depends critically on a large number of stakeholders, the majority of which are private landowners.

Objective 4-3. The example tactics are to communicate LCC successes and make LCC products more widely available. This objective is very similar in concept and practice to Objective 4-1.

TABLE 3.3 Brief Description of the LCC Network Objectives for the Third Goal, "Science"

LCC Network Goal- Objective #	LCC Network Objective Description
3-1	Identify shared science, information, and resource needs at the network scale.
3-2	Promote collaborative production of science and research—including human dimensions—as well as the use of experience and indigenous and traditional ecological knowledge among LCCs, Climate Science Centers, and other interested parties; use these to inform resource management decisions, educate local communities, and address shared needs.
3-3	Demonstrate and evaluate the value and improve the effectiveness of LCC science.

TABLE 3.4 Brief Description of the LCC Network Objectives for the Fourth Goal, "Communications"

LCC Network Goal-Objective #	LCC Network Objective Description
4-1	Communicate the existence and application of LCC Network science, products, and tools to partners and stakeholders in a form that is understandable, publicly accessible, engaging, and relates to what matters to end users and society.
4-2	Increase two-way communication with, outreach to, and engagement of key partners across the LCC Network as well as new partners to expand the LCC Network and increase conservation impact and achievements.
4-3	Develop and implement a communications and outreach plan that identifies and uses appropriate media to clearly convey to appropriate target audiences the value and tangible successes of the LCC Network at various scales.
4-4	Build communications capacity and capabilities within the LCC Network to effectively communicate the purposes and successes of the LCC Network.
4-5	Share lessons learned across the LCC Network.

Objective 4-4. This appears to be more of the same, basically an iteration of Objective 4-1.

Objective 4-5. This objective appears to belong under Goal 2, Collaboration. Clearly, one major advantage of having a network is to facilitate learning across wider time and space scales than would likely occur without a network; doing that is an aspect of collaboration.

TIMING OF EVALUATIONS IN THE PROGRAM LIFE CYCLE

The evaluation literature distinguishes between summative and formative evaluation (e.g., Wholey, 1996), both of which are retrospective evaluations of programs that have already been at least partly implemented. Summative evaluation takes place after a program has been completed or, in the case of ongoing programs, after a program has been fully implemented. The measurement criteria used for summative evaluation should be those associated with the ends and process objectives that appear in the objectives hierarchy, capturing the desired qualities the program is intended to enhance. However, for reasons of cost or convenience, proxy measures may be needed for some objectives, and those might be more closely associated with means objectives than with the ends that are fundamentally of interest. For example, if it is considered too difficult to directly measure the population status of a bird species dependent on large parcels of mature hardwoods, an evaluation might instead focus on the easier-to-measure extent and structure of the types of forest on which the species depends. Restoring forest habitat is a means to the desired end of enhancing the abundance of the bird, but because adequate habitat is a necessary, if not sufficient, condition for increased abundance, it can be used as a proxy measure.

Formative evaluation takes place partway through program implementation, or periodically during the course of an ongoing program, to assess progress and direct midcourse corrections. It is common for formative evaluation to use measurement of means as a proxy for accomplishment of ends because it may be too soon for even a successful program to have had its anticipated effect on ends. This is the case for the current evaluation of the LCCs, where the management of both natural and cultural resources that LCC activities are intended to influence has considerable inertia and may not show the effects of improved management for many years. In such circumstances, successful implementation of means, such as development of a nationally integrated land-use classification system, is an appropriate metric for formative evaluation.

Additional perspective on appropriate evaluation metrics for the LCC Network and for individual LCCs comes from the literature on the life cycle of governance structures for collaborative networks (Imperial et al., in press). These governance structures are viewed as having a natural evolution of both form and function. One model of this evolution lists four stages: (1) activation, where membership may be changing, relationships among members are developing, and the organization is searching for its focus; (2) collectivity, where the network has created processes for nurturing member interactions, has framed problems in a way that engages members and helps garner resources, and has codified decision-making procedures; (3) institutionalization, where the structure and role of the network have stabilized and the focus has turned to creating value through its expenditure of resources; and (4) stability, decline, or restoration, where a network may continue to garner resources and create value for some time, but may also need to adapt to address new problems or even disband, if the problems for which it was convened are now being addressed in another way. This is entirely consistent with focusing early evaluation on the means of achieving ends and on the processes by which decisions are made. In the longer term, when the LCC Network and individual LCCs have reached the institutionalization phase, it will be appropriate for evaluations to focus on creation of value as measured by achievement of the conservation ends for which the LCC Network was conceived. It is also worth remembering that reconfiguration or even dissolution can be part of the natural evolution of a collaborative network and that such an end casts no shadow on the value created by the network during its lifetime.

RELATIONSHIP OF THIS FRAMEWORK TO PERFORMANCE MEASUREMENT AND EVALUATION OF GOVERNMENT PROGRAMS

The federal agencies participating in LCCs are obliged to conduct performance measurement and program evaluation in the manner prescribed by the Government Performance and Results Act of 1993 (GPRA, 1993) and the Government Performance and Results Acts Modernization Act of 2010 (GRPA, 2010).[2] Chapter 4 critically examines an existing set of evaluation metrics used by the FWS to assess performance of the LCC system thus far (the Science Investment and Accountability Schedule); these metrics follow the Office of Management and Budget (OMB) guidelines. Chapter 4 also uses terms (e.g., process, outputs, outcomes) consistent with OMB guidance to propose new metrics for assessing LCC system performance in the short, medium, and long term. Because these terms are used commonly to discuss government programs, it seems worthwhile to note where the value-focused thinking framework used here, with its emphasis on distinguishing between ends and means, does and does not correspond to OMB terms.

OMB uses *outputs* to refer to products and services delivered by a program (see also Box 4.1); this terminology corresponds roughly to means in the value-focused thinking framework. OMB uses *outcomes* to refer to the results of those products and services; this corresponds roughly to ends in the value-focused thinking framework. The charge to this committee referred to outcomes and outputs consistent with OMB guidelines. In describing different types of program evaluation, OMB includes measurement of both outcomes and outputs as components of what it calls outcome evaluation. In contrast, value-focused thinking urges evaluation of programs with respect to their effects on ends, rather than with respect to their employment of particular means. However, value-focused thinking does recognize that means could appropriately be the focus of program evaluation, either as a proxy measure, where it is too difficult to directly measure accomplishment of ends, or as an interim measure, where it is too soon to expect to see any effect on ends, even for a successful program. OMB uses *process* to mean the type or level of activities conducted by the program, including adherence to regulatory requirements and professional standards. This is rather different from the value-focused thinking usage of process to refer to qualities of the way that a program undertakes its business (e.g., collaboratively).

CONCLUSIONS

In summary, similar to the language in the Secretarial Order, the strategic plan speaks mainly about means rather than about ends. The strategic plan does call attention to process objectives, as is appropriate for a program that intends to promote collaboration. The discussion above points out instances where goals appear to be means, rather than ends; where objectives seem to be misplaced under their respective goals; and where there are significant redundancies both among objectives under the same goal and among objectives that appear under different goals. We suggest reworking the language that describes the LCCs' overarching goals, vision, and mission with the aims to distinguish ends from means, to clarify process objectives that should be considered ends in themselves, and to elaborate sub-objectives. Such a reworking using the framework of value-focused thinking would lead clearly, and without redundancy, to evaluation metrics (see Chapter 4 on metrics). The same restructuring applied to the strategic plan would also facilitate development of succinct and operationally meaningful metrics for guiding implementation of the LCC system and evaluating its success. Upcoming revisions of the strategic plan offer opportunities for this restructuring.

The same advice about distinguishing ends and means, organizing ends objectives hierarchically, and using that hierarchy to develop an efficient set of metrics applies to individual LCCs and subgroups of LCC partners who are developing strategic plans under the LCC umbrella. These groups will be developing and revising strategic plans (including goals and objectives), choosing and implementing science and management activities, and evaluating their success in advancing the overall goals of the LCC system. Using a systematic approach to these recurring tasks will help make program evaluation more efficient and more transparent. It will also facilitate the essential task of relating the goals of the regional LCCs and their partner organizations to those of the national-level LCC Network.

Linking Individual LCCs' Goals with the Network's Goals

The strategic plan includes several goals and objectives that aim to facilitate coordination and collaboration across the individual LCCs and create a "high-functioning organizational culture for LCCs and the Network" (see above, Objectives 2-1, 2-2, and 2-3). Objective 2-3 of the LCC Network's goal—to develop conservation strategy—points to a need to improve the capacity for addressing broad-scale and cross-boundary problems. Identifying and addressing issues that span multiple LCCs or the entire network are unlikely to be addressed effectively through existing institutions and mechanisms. However, the network does not explicitly describe a process or structure that assists in identifying priorities at the larger or network scale. Nor are there metrics in place to aggregate outputs and results at the network scale (see Chapter 4 for further discussion). As discussed in Chapter 2, the landscape approach requires a bottom-up approach. The LCCs have each engaged in a strategic planning effort and begun to identify their strategic conservation priorities. As discussed in Chapter 4, each LCC is currently evaluated

[2] See https://www.whitehouse.gov/omb/mgmt-gpra/index-gpra and http://gao.gov/new.items/d11646sp.pdf.

using the Science Investment and Accountability Schedule evaluation tool, which is based on goals and objectives outlined in the Strategic Habitat Conservation Handbook (National Technical Assistance Team, 2008). The current difficulty in evaluating the LCC Network (discussed in Chapter 4 in greater detail) arises from an apparent lack in a process that clearly links the individual LCCs' strategic goals and missions to the LCC Network's strategic plan.

Many cross-LCC efforts are under way. However, developing an actionable strategic plan with clear national conservation priorities based on the conservation priorities of the individual LCCs has yet to be completed. The LCC Network staff have developed a table that illustrates how the goals of each LCC compare to the goals of the LCC Network Strategic Plan (see Appendix G). However, an evaluation process that can aggregate the accomplishments of individual LCCs to the network goals does not yet exist. To be able to do that, the evaluation tool for the individual LCCs has to change or the network's strategic goals/objectives have to more closely map to each LCC's goals (see additional discussion in Chapter 4).

The goals and objectives in the current strategic plan include most of the critical elements of the landscape approach discussed in Chapter 2. Overall, the strategic plan is consistent with the latest research on conservation at the landscape scale. The plan includes such critical elements as stakeholder engagement, adaptive management, and delivery of landscape-scale designs at the regional level with the aim to scale up at the LCC Network level. As discussed in Appendix C and elsewhere in this report, the guidance to implement the strategic plan needs to be improved. Developing science and information needs collaboratively with resource managers will be a critical process objective (and both an end in itself and a means of securing other ends), as discussed in Chapter 2.

Because the LCC Network was established in 2009, and some individual LCCs more recently, formative rather than summative evaluation is most appropriate at this stage. Formative evaluation can consider implementation of means (activities, outputs) as interim metrics of program success. Summative evaluation, concentrating on the effects that LCC activities have had on ends, will become more appropriate as the LCC program matures. As discussed in the section on timing of evaluations in the program life cycle, it is important for evaluation that the criteria being used to assess the health of the network during each phase in the program is tied to the structures and tasks appropriate to that stage. The LCC Network and its component LCCs are very much in the activation and collectivity stages of development, where evaluation is appropriately focused on engaging members; building relationships; and developing procedures for garnering resources, selecting projects, and carrying out activities.

Recommendation: When developing and revising strategic plans for the national network and for regional LCCs, the plans should distinguish ends, means, and process objectives, and organize ends objectives hierarchically to facilitate creating (and maintaining) an efficient set of evaluation metrics.

Recommendation: The LCC Network staff should conduct formative evaluation on an ongoing basis (e.g., annually) to guide LCC program implementation at regional and national levels.

Recommendation: The LCC Network should conduct summative evaluation periodically (perhaps every 5 years) for LCC programs that have been in existence long enough to have had a perceptible impact on ends (outcomes) to assess the values being enhanced by LCC activities.

4

An Examination of the Evaluation Process for the Landscape Conservation Cooperatives

This chapter discusses the rationale for designing an effective evaluation process. Subsequently, this chapter reviews the U.S. Fish and Wildlife Service's (FWS's) approach to evaluating the Landscape Conservation Cooperatives (LCCs) and provides guidance on what constitutes reasonable short-, medium-, and long-term metrics to evaluate the LCC Network. In this chapter, the committee responds to the following parts of the statement of task: What is the FWS's strategy to assess the effectiveness (outputs and outcomes) of the LCC Network? What are reasonable short-, medium-, and long-term metrics for the effectiveness of the LCC Network in achieving its stated purpose and goals?

The stated purpose or vision statement for the LCC program is to achieve "[l]andscapes capable of sustaining natural and cultural resources for current and future generations."[1] To achieve the prescribed vision and mission, the LCC Network has articulated its strategy (also described in Chapters 1 and 3), which contains four key goals: conservation strategy, collaborative conservation, science, and communications (LCC, 2014). Each goal is accompanied by a set of objectives, and each objective is further supported by "example tactics." Some of the tactics are essentially metrics used to measure progress in meeting individual goals for the network as a whole. The LCC Network Strategic Plan (hereafter referred to as the strategic plan; LCC, 2014) and its goals and objectives are described in Chapter 3, where we also offer a framework for setting clear means, process, and ends objectives. Here we review the FWS's evaluation approach and offer guidance to measure progress toward achieving the objectives of the LCCs and the LCC Network. This chapter begins with a discussion of the need for and challenges of program metrics.

[1] LCC Network Strategic Plan. Available at http://lccnetwork.org/strategic-plan.

NEED FOR AND CHALLENGE OF DEVELOPING PROGRAM METRICS

In general, and for the LCCs in particular, the challenges of establishing shared metrics and outcomes in a landscape-scale context (meaning multijurisdictional and multistakeholder) are well documented (see Box 4.1 for definitions). Typical challenges include, but are not limited to, the following (noting that some are explicit challenges to setting appropriate metrics, whereas others are challenges in process or outcome that may create difficulty both in setting appropriate metrics and/or later evaluating whether those metrics were correctly set):

- Defining an appropriate scale at which to measure conservation outcomes
- Unifying conservation objectives in situations where the leadership setting those objectives is distributed
- Integrating scientific information with management decisions
- Engaging multiple stakeholders, leading to diverse conservation objectives that may not work synergistically (e.g., Peterson et al., 2014)
- Considering the inherent complexity of the natural systems involved, in which apparent positive outcomes in any part of an ecosystem may involve trade-offs and/or create negative consequences in other parts of the ecosystem (NRC, 2005b; Game et al., 2013)
- Involving differences among focal questions, leading to varying methods of evaluation depending on the question being asked (Mascia et al., 2014)
- Distinguishing failure of a process or evaluation approach from failure of outcome (Peterson et al., 2014)
- Considering whether and under what conditions commonly espoused management techniques such as adaptive management are applicable (Allen and Gunderson, 2011; Doremus, 2011; NRC, 2011)

BOX 4.1
Definition of Key Terms

Metric is a standard of measurement by which efficiency, progress, performance, productivity, and/or quality of a deliverable, process, project, or product can be assessed.

Process metrics assess the type or level of program activities. (Note that this use of "process" differs from its use in the value-focused thinking framework presented in Chapter 3.)

Outcome metrics measure the results of products and services and, for the purposes of this analysis, correspond to measuring progress toward achieving ends objectives previously discussed in Chapter 3. Depending on the complexity and scale of the effort, outcome metrics might be required at multiple spatial and temporal scales.

Output metrics evaluate the products and services delivered by a program and, for the purpose of this analysis, correspond to measuring progress toward achieving means objectives previously discussed in Chapter 3.

SOURCES: Adapted from NRC, 2005c; GAO, 2011.

Given the presence of multiple stakeholder groups with multiple objectives and the absence of universally accepted approaches to monitoring and evaluating outcomes, it is not yet possible in a landscape-scale context to consistently evaluate whether the process, and associated metrics of the process, is directly linked to the outcomes later achieved and associated metrics of the outcome. In other words, establishing causal links between collaborative conservation planning and actions and the ultimate outcome is very difficult if not nearly impossible (such challenges are also described in Appendix A). Furthermore, because the LCCs primarily function as conveners and facilitators of collaborative conservation with the goal of "improving the management of fish, wildlife, and habitats," any evaluation will primarily demonstrate the LCCs' contribution to developing conservation strategies, delivering the science to inform management, and/or the quality of collaboration. Such evaluation challenges, however, are not unique to LCCs; nor does their recognition render irrelevant the need to articulate conservation measures for the purpose of outcome evaluation and program success. Indeed, a 2008 document prepared by the FWS titled *Strategic Habitat Conservation* describes the need for evaluating conservation programs in terms of inspiring "investor confidence."

APPROACH FOR EVALUATING CONSERVATION PROGRAMS USING PERFORMANCE MEASURES

Different approaches for evaluating conservation programs have been identified together with a set of focal questions to help determine the appropriate framework for evaluation (Mascia et al., 2014). Out of those, the *performance measurement* approach, which involves "the process of measuring progress toward specific project, program, or policy objectives, including desired levels of activities, outputs, and outcomes," appears most relevant here (Mascia et al., 2014). Its associated focal question is: "To what extent is a conservation intervention making progress toward its specified objectives for activities, outputs, and outcomes?" This approach is also consistent with the report titled *Performance Measurement and Evaluation: Definitions and Relationships* (GAO, 2011). It is important to note that this type of program performance evaluation can be effective in evaluating and improving conservation programs even without establishing a causal link between conservation actions of a program and the outcome in the health of a resource. Establishing such causal links is very difficult given the complexity of the system and the many uncontrolled variables.

Without suggesting specific individual performance measures, the Government Accountability Office (2011) identifies and suggests three *categories* of performance measures that are relevant across federal agencies: those that measure *process*, those that measure *outputs*, and those that measure *outcomes* (see Box 4.1). Furthermore, these performance measures meet the requirements set forth by the Office of Management and Budget (OMB). OMB developed guidance for program evaluations, in part to help agencies meet their requirements under the Government Performance and Results Act of 1993 (GPRA, 1993) and the Government Performance and Results Modernization Act (GPRA, 2010).[2] This law requires agencies to develop strategic plans and goals against which to measure performance. Annual reporting to OMB requires agencies to provide performance measures to assess outputs, service levels, and outcomes of each program activity as well as a comparison of accomplishments with performance goals set by agencies.

[2] See http://www.gpo.gov/fdsys/pkg/BILLS-111hr2142enr/pdf/BILLS-111hr2142enr.pdf.

U.S. FISH AND WILDLIFE SERVICE'S STRATEGY TO ASSESS THE INDIVIDUAL LCCs

Review of the FWS's Evaluation: Process, Outputs, and Outcomes Assessment of the SIAS Document

The FWS Science Investment and Accountability Schedule (SIAS) is the current approach of the FWS to assess the effectiveness of the LCC program. The SIAS was developed by FWS

> to express the Service's vision for, and to inform the Service's investment in, the suite of activities, actions and outcomes that an LCC would accomplish as it develops as a collaborative conservation forum; and, to help respond to Congressional direction that "the Service establish clear goals, objectives, and measurable outcomes for LCCs that can be used as a benchmarks of success in the program."

Benchmarks serve as a point of reference relative to which progress is measured (i.e., metrics). The SIAS consists of "eight, interrelated Conservation Activity Areas and twenty-two associated benchmarks that are guided by the Strategic Habitat Conservation Handbook (SHC; National Technical Assistance Team, 2008) in support of the LCC Network's Vision and Mission" (SIAS, 2013) (see Table 4.1). The foundation for the SIAS is the SHC handbook and several of the SIAS Conservation Activity Areas are directly identifiable in the SHC handbook (National Technical Assistance Team, 2008).

Each LCC coordinator completes the SIAS evaluation form for its respective LCC, which is designed to measure progress toward each of the eight SHC Conservation Activity Areas and their related progress benchmarks by means of associated metrics. An extract from the North Atlantic LCC SIAS 2.0 for 2014 (see Table 4.2) is provided to illustrate both the SIAS framework and a sample result from a completed evaluation form. The identification of benchmarks is important because they are reference points for measuring progress, and these benchmarks comply with the OMB requirement for performance goals.

Breaking the SIAS evaluation form down and reorganizing it according to the "process, outputs, and outcomes" categories in the Government Accountability Office report (GAO, 2011) enables a closer exploration of the organization of the SIAS document.

Challenges of Metric-Setting in the LCC Context

Since the development of the SIAS, the LCC Network has developed its strategic plan (LCC, 2014). Furthermore, the individual, self-directed LCCs have engaged in strategic planning efforts with their respective steering committees and many have developed their own strategic plans. Both the network-level strategic plan and the individual LCC plans tend to be written for internal, within-network audiences. It appears that they were not developed for the purpose of, and are not intended for, establishing metrics for program assessments.

TABLE 4.1 The 22 Benchmarks Under the Eight Conservation Activity Areas

Conservation Activity Area	Benchmarks
1. Organizational operations	1A. Engagement and coordination; 1B. Leveraging resources; 1C. Evaluating progress; and 1D. Engaged technical community and dedicated technical staff.
2. Landscape conservation planning foundation	2A. Assess existing conservation efforts; 2B. Identify priority resources; 2C. Collate and establish conservation goals and measurable objectives; and 2D. Refining landscape conservation planning foundation.
3. Landscape Conservation Design	3A. Vulnerability and landscape assessments; 3B. Adaptation strategies; and 3C. Integration of multiple priority resources and associated measurable objectives into Landscape Conservation Designs.
4. Informing conservation delivery	4A. Provide decision support; 4B. Information delivery; 4C. Assessment of information delivery; 4D. Collaborative conservation delivery to realize resource objectives; and 4E. Tracking delivery on the landscape.
5. Decision-based monitoring	5A. Collaborative monitoring; and 5B. Monitoring change of the landscape and priority resources.
6. Research to support adaptive management	6A. Testing underlying assumptions.
7. Data management and integration	7A. Data management and integration.
8. LCC Network function	8A. Participation in the LCC Network enterprise; and 8B. Function as part of integrated network of LCC partnerships.

TABLE 4.2 Science Investment and Accountability Schedule (SIAS 2.0) for the North Atlantic Landscape Conservation Cooperative (NALCC) for Fiscal Year 2014

SIAS 2.0 (FY 2014)			North Atlantic LCC		
Conservation Activity Areas and Benchmarks	Metrics	Metric Score	Bench-mark	Metric Score	Justification (limited to <4,000 characters)
1. Organizational Operations: Addresses fundamental organizational and administrative components necessary to establish and maintain an LCC as part of the national LCC Network. The LCC Partnership is composed of participating organizations (LCC Partners), is directed by the LCC Steering Committee (LCC SC), and is supported by the LCC Staff as well as science, technical, and other work teams. The LCC Staff and LCC SC and their associated organizations actively engage other relevant individuals, organizations, and partnerships creating collaborative relationships with key decision makers who are able to influence current and future landscape conditions. The LCC Staff maintains strong professional contacts and connections, networking to keep LCC Partners abreast of current conservation issues, techniques, etc. The LCC Staff also identifies partner capabilities to address the LCC mission and works with partners to address capacity gaps by adding key positions, relying on partner capacities, utilizing contracts, or by training appropriate to the size and complexity of the LCC geographic region (LCC Geography). LCCs must work closely with other conservation science and delivery activities to ensure efforts are coordinated and integrated. The LCC participates in development of common national LCC Network messages to relevant state, regional, and national entities. The LCC works to ensure its activities are coordinated and integrated with those of the Climate Science Centers, Cooperative Fish and Wildlife Research Units, Cooperative Ecosystem Studies Units, Forest Service Research Centers, Joint Ventures, Fish Habitat Partnerships, and similar key players.					

continued

TABLE 4.2 Continued

SIAS 2.0 (FY 2014)			North Atlantic LCC		
Conservation Activity Areas and Benchmarks	Metrics	Metric Score	Bench-mark	Metric Score	Justification (limited to <4,000 characters)
1.A - Engagement and Coordination - LCC Staff and Steering Committee are actively fostering strategic engagement, collaboration, and coordination with a diversity of entities that influence landscape conservation decisions, including state and federal agencies, tribes, universities, NGOs, regional partnerships (e.g., JVs, NFHPs, AFWA regions), and regional and local community planners.	No	0	1A	1	The North Atlantic LCC has a broad and active partnership of more than 100 partners including a Steering Committee with 33 members representing federal and state agencies, tribes, Canadian partners, and NGOs. Three technical teams and several ,project teams bring together technical expertise from agencies, universities, and organizations across the LCC geography. A science delivery team links to decision makers at regional, subregional, state, and local (land trust and community) scales. Regional partnerships including the Atlantic Coast Joint Venture, Eastern Brook Trout Joint Venture, Atlantic Coastal Fish Habitat Partnership, Northeast Partners in Amphibian and Reptile Conservation, Northeast Regional Ocean Council, Mid Atlantic Council on the Ocean, and others are linked to the LCC through team members and/or projects. The LCC has a particularly strong relationship with the Northeast Association of Fish and Wildlife Agencies at the director, administrator, and technical levels including a joint effort to support regional work in support of regional context for State Wildlife Action Plans.
	Yes	1			
1.B - Leveraging Resources - LCC Partners contribute resources (e.g., staff, funding, infrastructure, tools, expertise, etc.) to fill administrative and technical capacity, and information gaps necessary to achieve the LCC mission.	0% of total FWS annual investments leveraged by partner contributions (cash and/ or in kind).	0	1B	2	North Atlantic LCC partners contribute resources in numerous ways including staffing, in-kind participation and travel, complementary projects, and match. The U.S. EPA contributed a full-time liaison to the northeast LCCs (North Atlantic and Appalachian) through the first part of 2014 and the National Park Service contributes a portion of their coastal landscape adaptation coordinator's time to support LCC activities. The LCC and The Nature Conservancy shared a GIS analyst position to ensure spatial data are available to partners during part of 2014. About 50 non-FWS steering committee, technical team, and science delivery team members provide in-kind time and travel in support of LCC activities for several days a year. The close working relationship with the Northeast Climate Science Center results in a number of leveraged, complementary projects (e.g., integrated stream science projects). The NEAFWA RCN program provides directly complementary project support toward common goals in the Northeast Conservation Framework. Science Delivery partners use LCC funds to leverage their existing partner network capacity. The Connecticut River Landscape Conservation Design Pilot required significant in-kind participation by 30 FWS and non-FWS partners. The North Atlantic LCC successfully competed for > $5 million in DOI Hurricane Sandy Resiliency funds to address LCC priorities related to marsh, beach, and stream resiliency - about 5% of these funds were leveraged in 2014. LCC projects do not require match but several projects providing matching funds or in-kind services. The total of these contributions in 2014 was about 50% of the annual North Atlantic LCC budget.
	1% to 33% of total FWS annual investments leveraged by partner contributions (cash and/ or in kind).	1			
	34% to 66% of total FWS annual investments leveraged by partner contributions (cash and/ or in kind).	2			
	67% to 100% of total FWS annual investments leveraged by partner contributions (cash and/ or in kind).	3			
	>100% of total FWS annual investments leveraged by partner contributions (cash and/ or in kind).	4			

continued

TABLE 4.2 Continued

SIAS 2.0 (FY 2014)			North Atlantic LCC		
Conservation Activity Areas and Benchmarks	Metrics	Metric Score	Bench-mark	Metric Score	Justification (limited to <4,000 characters)
1.C - Evaluating Progress – The LCC Steering Committee has established metrics and processes for identifying, collaboratively pursuing, and evaluating actions in support of the LCC's mission, goals, and objectives. The LCC develops a comprehensive strategic action plan, updated on a regular defined time period, that describes their science agenda, approach, monitoring, and communications strategy and progress in collaboratively achieving the LCC mission.	*Part i: Has the LCC started a comprehensive strategic action plan?*				
	The LCC has not started a comprehensive strategic action plan.	0	1C(i)	2	The North Atlantic LCC completed a comprehensive Conservation Science Strategic Plan in 2011 as well as a draft science delivery plan and communications framework in 2013-2014.
	The LCC has started a comprehensive strategic action plan.	1			
	The LCC has completed a comprehensive strategic action plan.	2			
	Part ii: Has the LCC Steering Committee started a process for evaluating progress?				
	The LCC Steering Committee has not started a process for evaluating progress at regular intervals toward established goals and updating the identification and prioritization of the most important science and capacity needs to support LCC goals.	0	1C(ii)	2	LCC staff provide a state of the LCC presentation annually at each April Steering Committee meeting and the Steering Committee reviews and provides input on shifting of priorities. In 2013–2014 the Steering Committee recommended a shifting of resources toward science delivery that was reflected in a new team, strategy, and grant program for science delivery. The LCC technical teams review science priorities each year and provide recommendations on needs to address and update the science needs matrix portion of the strategic plan. In 2013–2014, that review resulted in adaptive actions to fund work that would be responsive to and complementary to Hurricane Sandy resiliency funded work and ongoing Landscape Conservation Design projects. In 2014, the Steering Committee and staff initiated a review of the strategic plans with the intent of consolidating and updating them in 2015–2016.
	The LCC Steering Committee has started this process	1			
	At least one iteration of this process, resulting in an updated strategic action plan, has been completed. *Note: Report (in narrative form) on the identified adaptive actions taken as a result of the process.*	2			
	1.C Summary Score		1C	4	
1.D - Engaged Technical Community and Dedicated Technical Staff - The LCC has organized the technical capacity, including dedicated partner staff, needed to address priority conservation science needs. Further, the LCC has established a working relationship with USGS regional Climate Science Center(s) and other entities to ensure that science and conservation activities involving the LCCs have access to the best regional technical information and that priorities are coordinated and integrated.	The LCC has not organized technical capacity nor established relationships with the broader science community.	0	1D	2	The North Atlantic LCC has three engaged technical teams addressing coastal and marine, terrestrial, and wetland and aquatic science needs in the LCC science strategic plan. The LCC has a science coordinator, science delivery coordinator, data manager, and GIS analyst to provide technical staff capacity and staff support through EPA and NPS liaisons and a shared position with TNC. The North Atlantic LCC has strong working relationships with both the university and USGS components of the Northeast Climate Science Center (located 1 mile from the LCC office) and has been directly involved in developing and ranking the results of all CSC RFPs. Project oversight teams and peer reviewers ensure the LCC projects achieve their stated goals.
	The LCC has established science teams or technical committees to assess science and technical needs for the LCC.	1			
	The LCC's science teams or technical committees are addressing the LCC's priority conservation science needs.	2			

NOTES: Shown is an extract of the SIAS 2.0 table for the NALCC with a description of the Activity Areas to be evaluated, titled "Organizational Operations" with corresponding benchmarks, metrics, and options of metric scores in the gray, left side of the table. On the blue-colored, right side of the table, the NALCC assigned itself a metric score and provided a justification for the metric score. The yellow highlighted area indicates the score the NALCC assigned to each benchmark. AFWA = Association of Fish and Wildlife Agencies; CSC = Climate Science Center; EPA = U.S. Environmental Protection Agency; FWS = U.S. Fish and Wildlife Service; FY = Fiscal Year; GIS = geographic information system; JV = Joint Venture; NEAFWA = Northeast Association of Fish & Wildlife Agencies; NFHP = National Fish Habitat Partnership; NGO = nongovernmental organization; NPS = National Park Service; RFP = request for proposals; TNC = The Nature Conservancy.
SOURCE: FWS.

The SIAS is currently the only evaluation tool being used by the FWS and generally is focused on FWS-centric metrics, but, as the committee learned during discussions with LCC staff, it is not necessarily adaptable to the network and individual LCCs from the broader perspective necessary to measure complex landscape- and partner-driven outcomes. A single SIAS or sets of SIAS-focused metrics that encompass all important program elements will be complicated and perhaps dampen innovation in design. For this reason, the SIAS framework can meet objectives of the FWS independently from an evaluation framework that assesses the LCC Network and individual LCCs as long as some consistency is maintained at the network level. The discussions in this chapter do not assume the SIAS is the only metrics option, and indeed the LCC Network may wish to consider developing a separate framework outside of the SIAS.

Reflective of the challenges noted above, the structure of the LCC Network poses some unique challenges with regard to setting metrics. In most contexts, the focus of program assessment and metric-setting is to demonstrate how a respective agency's investments have resulted in achieving the particular agency's goals. In the LCC Network, where goals are to be achieved through collaborations across programs and agencies, subsequent outcomes must also be viewed in the context of these partnerships. This challenge is acknowledged by the FWS in its discussion of the SIAS (2013) and includes the recognition that additional measures must be forthcoming, stating that:

> The partnership of LCCs provides the opportunity for significant conservation progress, but also demands a complex set of interactions in which different agency and group authorities and priorities must be respected. This version of the SIAS is referred to as SIAS 2.0 in recognition that it is the next evolution of the original SIAS performance system. . . . SIAS 2.0 is another step along the path to develop an incentive based approach to Service investments in LCCs. Therefore, SIAS 2.0 will require further refinement after it is implemented and eventually a new SIAS 3.0 will result.

This statement also acknowledges the intent to carefully use metrics that reflect only FWS effectiveness in implementing the network of LCCs, and not the outcomes that should result from the collaborative process.

DEVELOPING APPROPRIATE SHORT-, MEDIUM-, AND LONG-TERM METRICS

Appropriate Short-Term Metrics

One of the LCC Network's greatest strengths is its ability to respond to stakeholder priorities within a given region and its flexibility to collaborate across LCCs in response to conservation challenges that are multijurisdictional. However, this characteristic can also be a weakness in that it can create the appearance of a program serving a diversity of missions and objectives, and lacking focus. During a discussion with the committee, it was noted that the diversity of LCCs has led to a bewildering diversity of LCC science "requirements," formats, priorities, and communication mechanisms. In general, the LCC Network needs to be able to demonstrate that it operates both efficiently and as a coherent program, and that the program's constituent elements work together toward the program's overall stated purpose and objectives and do so cost-effectively. For this reason, most of the short-term metrics listed below are process oriented, and are intended to refine FWS-relevant benchmarks associated with the SIAS, as well as to strengthen and clarify alignment of the SIAS with the strategic plan.

- *Extent to which costs for each SIAS Activity Area can be identified and returns on investment relative to those costs identified.* Because LCCs work through partnering with relevant agencies and much of the implementation authority lies outside the LCCs, it is important that they can demonstrate how investments were either leveraged or resulted in outputs or outcomes through activities or actions undertaken by LCC partners. Furthermore, because facilitating the development of shared conservation priorities is an important component, it is important that the SIAS also evaluate the breadth and scope of partner engagement (ACCCNRS, 2015). At present, the SIAS 2.0 contains a section titled "Leveraging Resources," which is the only benchmark that identifies resources as an issue. This benchmark defines the contribution of the total FWS annual investments that have been leveraged by partner contributions, either cash, in kind, or both, under the Activity Area of "Organizational Operations." For example, to reduce cost of office space, LCC coordinators for two LCCs work in offices provided by state agencies. Additionally, in the case of the Desert LCC, the Bureau of Reclamation provides the staff support to coordinate the LCC. Many LCCs receive technical and staff support from other state or federal agency partners on the steering committees. To the extent possible, it would be appropriate for the next iteration of the SIAS to more fully describe—either qualitatively or quantitatively—the returns achieved by the LCC Network for both the FWS and partner dollars invested.
- *Extent to which benchmarks are appropriately assigned and distributed to achieve progress in each SIAS Conservation Activity Area.* In the SIAS, benchmarks are more fully assigned to some Activity Areas than to others. For instance, "Landscape Conservation Design" contains numerous benchmarks addressing a wide range of strategies and services, while "Research to Support Adaptive Management" contains one benchmark. This uneven distribution of benchmarks could be interpreted as assigning value to the various Activity Areas and may or may not address fully the intent of that Activity Area. Although the benchmarks identified in some of the Conservation Activity Areas are appropriate, consideration might be given to whether the number of benchmarks needs to be evened out and whether

additional benchmarks may be needed in other areas to better define and measure progress.

• *Degree to which SIAS benchmarks are associated with the appropriate short-, medium-, and long-range context.* Because the SIAS progress benchmarks are not couched in the context of short-, medium-, and long-term progress, the accompanying metrics are also not framed in that context. Not all progress toward meeting benchmarks could proceed at the same pace or not all outcomes could be achieved at the same time.

• *Extent of alignment at the network level between the SIAS and other key documents, i.e., the strategic plan.* When developing program metrics, there usually is a fundamental framework that the metrics address. The basic framework of the LCC Network is the strategic plan. The strategic plan is based on a vision and mission with four strategic goal areas: (1) conservation strategy, (2) collaborative conservation, (3) science, and (4) communications. Nineteen objectives provide direction on how these goals will be accomplished (see also Chapter 3). When attempting to cross-reference the goals and objectives of the strategic plan to the metrics of the SIAS, a direct match of metrics to goals and objectives is not obvious. Many of the 22 benchmarks can be paired to single or multiple goals and objectives, but it is certainly not a set of metrics designed to align with the strategic plan—in part, because the SIAS uses the SHC framework for metrics while the strategic plan is not based on the SHC framework directly; and also in part because this version of the SIAS was completed before strategic plan development. This sequence of events could lead to some confusion and difficulty in communicating LCC effectiveness.

Assigning Appropriate Medium- and Long-Term Metrics

What constitutes medium and long term is a matter of judgment. Generally speaking, conservation outcomes are context dependent, and what might be appropriate as a medium-term metric in one situation or scale might be more appropriate as a short- or long-term metric in another. Nonetheless, with the intent of defining program phases, since the LCC Network was initiated 5 years ago (in 2010), it would be appropriate to assign a working definition to short term of 0–2 years from today (or 7 years since program inception), to classify medium term as 7–10 years from present, and to refer to long term as more than 10 years from present. This classification acknowledges that it takes a long time to achieve ends objectives (see discussion in Chapter 3), and that measuring some of the outcomes of a program might take much longer than the next 10 years, whereas others will not.

Using the process, output, and outcome assessment approach identified above, it will be helpful to conceptualize medium- and long-term metrics in terms of those that measure the outputs delivered and the outcomes they help achieve as well as those that measure the effectiveness of the collaborative process itself (see also discussion in Chapter 3 on ends and means objectives). As the LCC Network evolves, the SIAS metrics and those of any other LCC evaluation framework that might be developed need to be refined to more fully address the following factors.

Measuring Outputs and Outcomes: Outputs and outcomes are not needed for every action that an LCC takes; however, the LCC Network needs to be able to identify accomplishments that have occurred as a result of the LCC Network being in place, and also needs to be able to easily identify one or more clear successes for individual LCCs. Thus, in the committee's assessment, measuring the medium- and long-term effectiveness of the LCC Network needs to include the extent to which funds can be associated with each of the SIAS Activity Areas (see Conclusion section). Measuring effectiveness also needs to be tied to appropriate medium-term outputs (products and services delivered) and longer-term outcomes (results achieved for the products and services delivered).

From this standpoint, reasonable medium- and long-term metrics for individual LCCs, which could then be aggregated at the network level, would assess the following:

• *Extent to which both agency-unique contributions and their costs can be clearly identified and tied—qualitatively, quantitatively, or both—to program outputs and longer-term outcomes.* As the LCC Network evolves, and in light of increasing emphasis on government performance and results, it is important for each of the public agencies involved to attempt, to the extent feasible and practicable, to identify its unique contribution and the costs of that contribution to the collective impacts achieved by the LCC Network. Metrics ideally include results achieved for both financial and human capital investments.

• *Extent to which partner contributions and the costs of those contributions can be clearly identified and tied—qualitatively, quantitatively or both—to both outputs and outcomes.*

• *Extent to which both successes and failures can be meaningfully analyzed in ways that result in program corrections.* As Baylis et al. (2015) wrote, "Understanding why conservation programs succeed or fail is essential for designing cost-effective initiatives and for improving the livelihoods of natural resource users."

Measuring Collaboration: In addition, medium- and long-term metrics also need to measure the outputs and outcomes delivered by the collaborative process itself. Because the main goal of the Secretarial Order was to create a mechanism for facilitating collaborations across jurisdictional boundaries, it is important to measure the quality of and the deliverables from the collaborative process. As discussed in Chapter 3, this means objective is, at the same time, an ends objective. However, measuring collaborations and outcomes from these collaborations is difficult for several reasons, as discussed above: (1) the FWS needs to meet the expectations from both OMB and Congress and demonstrate outcomes directly relevant to its core mission; and (2) measuring collaborations at the individual LCC level cannot be easily aggregated to demonstrate effectiveness of collaboration at the network level. Therefore, the committee provides a review and guidance on this measurement topic.

The work of Thomson et al. (2007) may be helpful to the LCC Network on an illustrative basis, because it identifies five key dimensions contributing to an overall construct of collaboration. Thomson et al. (2007) define collaboration as "a process in which autonomous or semi-autonomous actors interact through formal and informal negotiation, jointly creating rules and structures governing their relationships and ways to act or decide on the issues that brought them together; it is a process involving shared norms and mutually beneficial interactions."

The authors suggest five factors relevant to measuring collaboration, briefly recited below. In seeking empirical data to measure the model's validity, the authors structured survey questions that were sent to 1,382 managers of a large national organization; the survey approach may be something the LCC Network would want to consider in order to gather feedback on the effectiveness of the collaboration process. The collaboration metrics associated with each of the five factors could be rated using the format currently applied in the SIAS document, that is, a progressive, scaled rating system consistent with the current SIAS approach to measurement. The five factors are shown below, together with some suggested metrics that the committee finds relevant to the LCC Network in terms of measuring the effectiveness of its own collaborations.

- **Governance** involves "developing sets of working rules about who is eligible to make decisions, which actions are allowed or constrained, what information needs to be provided, and how costs and benefits are to be distributed" (Ostrom, 1990). Illustrative metrics, appropriate to LCC Network governance could include (a) identifying where overlapping objectives exist, and if such overlaps exist, (b) identifying the extent to which policy or procedural platforms exist that allow one organization to take a lead for a given issue to allow interagency collaboration to continue.
- **Administration** means that "some kind of administrative structure must exist that moves from governance to action . . . as public mangers [sic] know all too well, decentralized administrative structures still require a central position for coordinating communication, organizing and disseminating information, and keeping partners alert to the jointly determined rules for governing relationships—what Freitag and Winkler (2001, p. 68) describe as social coordination." Illustrative metrics appropriate to administration could include (a) whether there is an administrative structure in place to allow agencies to identify who will coordinate, communicate, and disseminate information throughout the network, and (b) the extent of its effectiveness.
- **Organization autonomy** recognizes and acknowledges that collaborating organizations "maintain their own distinct identities and organizational authority separate from a collaborative identity . . . when collaboration's goals conflict with the autonomous goals of individual partner organizations, identities are at stake . . . it is likely that individual missions will trump collaboration missions" (Thomson and Perry, 2010, p. 293). Because the individual LCCs set their own goals and priorities through a collaborative process (i.e., the steering committees), each LCC develops a distinct identity based on the joint goals of the partners. Therefore, the metrics appropriate for each individual LCC may or may not be identical. Furthermore, metrics for individual LCCs ideally assess the contributions of partners to the individual LCC's goals, and vice versa. Thus, illustrative metrics appropriate for the LCC Network could include metrics to measure the contributions of partners.
- **Mutuality** is closely tied to complementarity and describes a situation where "parties to a network agree to forgo the right to pursue their own interests at the expense of others" and accommodation serves as the modus operandi of interaction (Powell, 1990, p. 303). "It occurs when one party has unique resources (e.g., skills, expertise, and money) that another party needs or could benefit from (and vice versa). As long as collaborative partners can satisfy one another's differing interests without hurting themselves, collaboration can occur" (Wood and Gray, 1991, p. 161). Illustrative metrics appropriate for the LCC Network could include the extent to which LCC Network partners have been able to leverage contributions made by other partners in the network.
- **Trust** is a central component of collaboration because it reduces complexity of transactions. Developing trust takes time and needs repeated interaction among partners to build the credible commitment necessary for collective action to occur (Axelrod, 1984, 1997; Ostrom, 1990). The Comprehensive Everglades Restoration Plan and its implementation—and to some extent, the lack of its implementation—is a study in trust-based collaboration (NRC, 2014b). A challenging restoration effort is taking place in the California Bay-Delta, which also involves complex institutional arrangements requiring trust building (NRC, 2012b). Illustrative metrics appropriate for the LCC Network—reciprocity and trust—could include the extent to which trust among parties has led to reduced complexity of transactions.

Establish and Aggregate Individual LCC Metrics to Measure LCC Network-Level Goals

In presentations to the committee and public, LCC staff acknowledged difficulties scaling up from individual LCC SIASs to a network-level SIAS assessment. In addition, other than completion of the SIAS by each LCC coordinator, most LCCs have yet to develop their own set of metrics beyond the SIAS, although many have some requirements for measuring success. For example, the strategic plans for the Eastern Tallgrass Prairie and Big Rivers LCC, the Aleutian and Bering Sea Islands LCC, and the North Pacific LCC have expressed the intent to develop metrics that mimic either the SIAS or some portions of the SIAS. In open discussion with individual LCC participants, it was clear that while metrics for individual LCCs have only been developed by a few LCCs to date, the need for such metrics is generally acknowledged as a priority by most LCCs. Annual reports include benchmarks per se, but those reviewed provided accomplishments without first identifying the need for the accomplishment to occur or including metrics to evaluate progress toward those accomplishments.

The FWS staff has mapped the individual LCCs' goals onto the goals of the LCC Network Strategic Plan (see Appendix G). This illustrates the complexity of developing metrics for individual LCCs that can assess the effectiveness of the LCC Network as a whole. It further demonstrates the need for compatible metrics that cross all LCCs and provide some level of outcome measurement at the LCC Network level.

CONCLUSIONS AND RECOMMENDATIONS

The committee concludes that the SIAS document provides a useful initial assessment of the FWS components of the LCC Network. Specifically, (1) the categories of performance metrics (process, outputs, and outcomes) outlined by OMB were addressed (although not labeled as such); (2) each of the eight Conservation Activity Areas guided by the SHC framework establishes progress benchmarks; (3) metrics were linked to the progress benchmarks in the SHC framework; (4) benchmarks and their accompanying measures were set at a level of aggregation that appropriately transcends the conservation issues specific to any one LCC (or subset of LCCs) and provides a common system for programmatic evaluation that is network wide; and (5) some of the metrics used a "percent complete" approach, whereas others were expressed narratively; however, the context for each was appropriate. Thus, the committee concludes that the SIAS helps the FWS meet its reporting requirement to Congress and OMB.

The committee concludes that the current approach and focus on measuring FWS investments and outcomes do not adequately measure all aspects of the LCCs and the LCC Network. In particular, the committee concludes that the evaluation process currently falls short in two important ways: SIAS metrics are not aligned with the goals of the LCC Network or with those of the individual LCCs; and the SIAS does not measure the value of the network to its partners.

Aligning SIAS Metrics and Benchmarks with Network Goals: Because the SIAS benchmarks are based on the goals in the FWS's SHC plan, the SIAS is not designed to measure progress toward the goals in the LCC Network Strategic Plan. Also, some SIAS metrics account for efforts across multiple LCCs, but a process is missing to effectively measure various efforts at the network scale, or to aggregate results from the individual SIAS documents to the network scale. To improve the SIAS, an important short-term need will be the alignment of the next iteration of the SIAS to better reflect the mission, goals, and objectives of the strategic plan, as well as the goals and objectives of each self-directed LCC.

As discussed in Chapter 3, there are opportunities to improve upon the strategic plan, and based on discussions with FWS staff, the committee expects that a new iteration of the LCC Network plan will be forthcoming. In doing so, the FWS has to consider whether the SIAS can assess the goals of the LCCs and the LCC Network, while still serving the intended purpose of the SIAS, which states,

> the FWS [(SIAS)] will help guide our support for individual Landscape Conservation Cooperatives (LCCs) and the National Landscape Conservation Cooperative Network. In pursuit of our agency's mission and our vision for science, the following Activity Areas and associated benchmarks will help specify our investment and participation in the LCC Network to ensure effectiveness, efficiency, and support for the LCC Network vision and mission.[3]

We conclude that establishing solid and defensible metrics that are clearly aligned to processes, outputs, and outcomes at the individual LCC level is a key step toward creating a more understandable and defensible LCC Network as a whole. For that reason, it is important to direct LCC resources into developing sound metrics at the individual LCC level. The cost (staff time) required to complete such evaluations needs to be considered in the development of further evaluation processes and tools.

Recommendation: The FWS, in its next iteration of the SIAS, should (1) identify how and where the SIAS relates to elements of the LCC Network strategy; (2) identify (as it has done for SIAS 2.0) the benchmarks associated with each Activity Area, and continue that exercise by (a) classifying benchmarks as short, medium, or long term; (b) ensuring that benchmarks are adequately developed for and assigned to each SIAS Activity Area; and (3) begin the process of identifying, to the extent feasible and practicable, costs relative to returns on investment associated with achieving each benchmark.

Recommendation: Establishment of metrics at the individual and network-wide scales should become a high priority.

- Metrics should be developed to measure each LCC's unique goals, yet be consistent enough across LCCs to permit aggregation to a network scale.

---

[3] See http://www.fws.gov/science/pdf/SIAS-FY2013.pdf.

• The criteria used to rate the performance of the LCC Network as a whole, and its components, should be closely related to the objectives that they are intended to evaluate, and articulated clearly enough that any evaluator with access to the same information about the LCC Network could apply those criteria consistently.

• To more clearly demonstrate relevance to the stated purpose and goals of the LCC Network, as well as to better define the FWS role in support thereof, the SIAS Activity Areas and benchmarks should be written in a manner that clearly aligns with the LCC Network's purpose and goals, as captured in the strategic plan or its next iteration.

Measuring Partner Contributions and Benefits: Although the SIAS tracks how partners contribute to the FWS-specific goals of the individual LCCs, currently the evaluation process does not yet account for how these contributions further the goals of the network in general or the goals of the key partners in particular. Thus, the current evaluation approach will not be able to account for the outcomes that result from convening partners, and it might fail to properly measure partners' investment in or benefits from joint activities. As a result, it might lead to misinterpretation of accomplishments and/or an inability to express the value-added outcomes of the LCCs. Developing metrics for the network as a whole may best be incorporated into a frame of reference that is complementary to, but separate from, the SIAS. Because the SIAS is an FWS product that is reflective of the agency's own goals and objectives, the LCC Network as whole may consider a framework that is complementary with, but separate from, the SIAS in order to better capture goals, objectives, and measurements toward network-wide progress. In so doing, the observations in the above section on short-, medium-, and long-term metrics will be useful and applicable. In short, the requirement for short-term "process" metrics never disappears—an organized, coherent program is necessary to justify government funding, and the process organization should carry through and become increasingly refined as the program evolves. In this sense, monitoring the extent of and continuing to strive for alignment between the SIAS and other key documents, such as the strategic plan and the individual strategic plans of the respective LCCs, should continue. Furthermore, the process of better organizing the SIAS, using the short-term metrics identified above, will likely give rise to additional medium- and long-term metrics beyond those identified below. While providing a deliberate, useful, and meaningful initial assessment tool, the SIAS document and its accompanying metrics would benefit from supplementing in ways that would refine the use of metrics to better show the returns achieved for investments made by partners and design and apply FWS-relevant metrics at the individual LCC level that can be aggregated to measure and demonstrate outcomes at the LCC Network level.

Recommendation: The LCC Network should improve its evaluation process to better capture the contributions made by all partner agencies or groups toward common objectives. In particular, to demonstrate the effectiveness of the individual LCCs and the LCC Network, the evaluation process should measure how resources invested in any portion of the LCC Network further the goals of the LCC Network and its partners. The efforts invested in the LCCs and the LCC Network consists of (1) federal funding allocation via the FWS; (2) partners' in-kind contributions via staff time or technical expertise; and (3) funding from other state/federal agencies or private partners.

In particular, the committee recommends:

1. Measure contributions and results—distinguish among process, output, and outcome metrics—and attempt to ensure that the outputs (the products and/or services rendered) and the outcomes (results achieved for those products and services) are clearly identified where feasible to do so. Specifically,
    a. seek to identify agency-unique contributions as well as partner contributions and the outputs/outcomes achieved for those contributions;
    b. seek to identify costs of agency-unique and/or partner contributions relative to the outputs/outcomes obtained for those investments; and
    c. continue to use both qualitative and quantitative measures to evaluate outputs/outcomes.
2. Where feasible, measure collaboration, that is, governance, administration, organizational autonomy, mutuality, reciprocity, and trust.
3. Where feasible, develop impact and collaboration metrics at the individual LCC level in ways that aggregate to the network level.

5

The Landscape Conservation Cooperatives and Other Similar Federal Programs

In this chapter, the committee addresses Tasks 2 and 3 of the statement of task (see Chapter 1), which, in brief, ask the committee to (a) compare the Landscape Conservation Cooperatives (LCC) program with other similar programs, considering similarities, differences, overlap, and coordination; and (b) compare activities supported by LCCs and related programs within the U.S. Fish and Wildlife Service (FWS) and other agencies, and whether there is sufficient coordination and integration. To address these tasks, the committee reviewed and discussed similar programs that are federally funded[1] (see Table 5.1 for a brief overview). There is a large number of existing federal programs that focus on conservation, and this review should not be considered comprehensive. Rather, this chapter and Appendix D include programs reviewed by the committee and, in particular, programs that were commonly referred to as potentially overlapping with the LCCs during the committee's information-gathering efforts.

The committee considered initially more than 20 programs, including federal research laboratories and U.S. Geological Survey (USGS) Cooperative Research Units. After cursory review, the committee narrowed its analysis to 11 other programs that operate sub-nationally but across a considerable span of the United States and affiliated territories (in some cases, as with LCCs, these groups also have regionally organized governance structures), and that have a focus on landscape conservation and/or adaptation to climate change. The committee gave no further consideration to programs, including those within the U.S. Department of the Interior (DOI), whose primary purpose is the generation of scientific research, but that otherwise lack multiple key characteristics in common with LCCs, such as a landscape-scale focus, stakeholder involvement, or a large geographic domain.

[1] Because of the vast number of collaborative conservation efforts throughout the nation, the committee elected to interpret "similar programs" and "related programs" as those within the purview of federal agencies.

Considering the five attributes below (see Box 5.1), the committee examined the 11 programs listed in Table 5.1 for overlap with the LCCs, and identified four programs sufficiently similar to LCCs in scope, scale, organization, and emphasis to warrant a closer examination: FWS's Migratory Bird Joint Ventures program (Joint Ventures), FWS's Fish Habitat Partnerships (FHPs), USGS's Climate Science Centers (CSCs), and the National Oceanic and Atmospheric Administration's (NOAA's) Regional Integrated Sciences and Assessments (RISA) program. A brief discussion of the other seven programs is provided in Appendix D. The committee also discusses the regional-scale coordination among programs in the Pacific Islands and the Southeast United States. Finally, the committee concludes the chapter with feedback on how the LCCs can be best positioned to support and coordinate with similar conservation efforts or related activities.

DETAILED CONSIDERATION OF SIMILAR PROGRAMS

To gauge the programs' similarity to LCCs, the committee considered the five attributes listed in Box 5.1. After examining numerous documents describing the various programs and considering these five attributes, the committee judged the degree of similarity to the LCCs to be roughly in the order presented in Table 5.2 and chose to concentrate its in-depth analysis on FWS's Joint Ventures, FWS's FHPs, USGS's CSCs, and NOAA's RISA Program.

Migratory Bird Joint Ventures

Responding to reductions in waterfowl populations and habitat, the U.S. and Canadian governments developed the North American Waterfowl Management Plan (NAWMP, 2012) to restore waterfowl populations. The NAWMP was first adopted in 1986 and has been updated several times since then. Congress passed the North American Wetlands

TABLE 5.1 Overview of the LCCs and 11 Other Federal Programs Considered[a]

Entity and Primary Focus	Geography, Structure, and Governance	Mission (Paraphrased)
Landscape Conservation Cooperatives (LCCs)* Conservation strategies through technical support	22 LCCs cover the U.S. land area, territories, Pacific and Caribbean islands, and parts of Canada and Mexico, and broadly participate in many activities. They are advised by a steering committee with members from a range of public and private conservation and resource management partners.	LCCs develop and provide integrated science-based information about the implications of climate change and other stressors for the sustainability of natural and cultural resources and develop landscape-level conservation objectives.
Migratory Bird Joint Ventures (JVs)* Large habitat or species conservation	22 habitat-based JVs and three species-based JVs cover all of the United States, Canada, and a large part of Mexico. They are advised by an independent management board with members from a range of public and private conservation and resource management partners.	JVs benefit migratory bird populations, other wildlife, and the public by sustaining a diversity of habitats through cutting-edge science and technology.
U.S. Fish and Wildlife Service (FWS) Fish Habitat Partnerships (FHPs)* Large habitat or species conservation	19 FHPs (18 based on particular geographies or species, and 1 based on a particular type of system) cover the U.S. land area, territories, and parts of Canada.	FHPs protect, restore, and enhance fish and aquatic communities and habitats. The partnerships foster fish habitat conservation and improve the quality of life for the American people.
U.S. Geological Survey (USGS) Climate Science Centers (CSCs) and the National Climate Change and Wildlife Science Center (NCCWSC)* Research to support management of fish, wildlife, and habitat	Eight centers cover the U.S. land area and territories. Each is a USGS–university cooperative agreement and has an advisory council composed exclusively of federal, state, and tribal representatives. NCCWSC provides overall coordination for the eight CSCs.	CSCs provide science support to natural resource managers for dealing with effects of climate and other concurrent global changes on fish and wildlife and their habitats.
National Oceanic and Atmospheric Administration (NOAA) Regional Integrated Sciences and Assessments (RISAs) Build resilience to climate through stakeholder-driven research	11 RISAs,[b] all housed at universities (domain determined by proposers), collectively cover roughly 75% of U.S. land area. Priorities, topics, and governance vary widely.	RISAs act as a research engine for partnership-driven science to expand the nation's capacity to prepare for and adapt to climate variability and change.
National Park Service (NPS) Scaling-Up initiative* Conservation of areas adjacent to national parks	The initiative focuses on the 408 units of the national park system (covering 84 million acres) operated by the NPS, including 30 national historic and scenic trails and 49 National Heritage Areas.	The initiative preserves natural and cultural resources within national parks, trails, heritage areas, and landmarks by improving conditions beyond those boundaries through a larger-landscape approach to conservation.
Bureau of Land Management (BLM) Landscape-Scale Approach to Managing Public Lands* Resource conservation, restoration, and development	BLM administers more than 245 million acres located in the 12 western states. BLM Ecoregions cross traditional administrative boundaries.	The ecoregional approach identifies important habitats for fish, wildlife, and species of concern, and their vulnerability to climate change, wildfires, invasive species, and development.
NOAA Regional Collaboration Teams Conservation and restoration of coastal and marine habitat	Eight Regional Collaboration Teams within NOAA integrate NOAA employees and affiliates to promote regional coordination of NOAA assets in order to better address stakeholder concerns.	The Regional Collaboration Teams serve as flexible networks of NOAA staff and affiliates who engage with stakeholders, assess the needs of external partners and stakeholders, and adjust NOAA products and services accordingly.
U.S. Department of Agriculture (USDA) Regional Climate Hubs Provision of climate information to private landowners	Seven Regional Climate Hubs cover the U.S. land area and some territories. The program currently consists primarily of part-time directors with small budgets.	Regional Climate Hubs provide information to farmers, ranchers, and forest landowners to help them adapt to the impacts of climate change and promote sustainability.

TABLE 5.1 Continued

Entity and Primary Focus	Geography, Structure, and Governance	Mission (Paraphrased)
U.S. Forest Service (USFS) Collaborative Forest Landscape Restoration Program (CFLRP) All lands, collaborative, science-based ecosystem restoration of priority forest landscapes	USFS manages more than 192 million acres of forestlands and grasslands. The CFLRP is an approach to conserving priority forests led by USFS in close coordination with other landowners to encourage collaborative solutions through landscape-scale operations. The program uses a competitive process to allocate funding to collaborative groups to implement management activities.	The CFLRP encourages ecological, economic, and social sustainability, reduces wildfire management costs, and demonstrates effectiveness of ecological restoration techniques.
Natural Resources Conservation Service (NRCS) Landscape Conservation Initiatives Helps the agricultural sector contribute to conservation goals best addressed on a landscape scale	Nationwide program with designated priority areas—Critical Conservation Areas. Also includes Chesapeake Bay Watershed and the Mississippi River Basin. It provides assistance to landowners through grants from the Regional Conservation Partnership Program (RCPP).	RCPP promotes coordination for landscape-scale initiatives among NRCS and its partners to assist producers and landowners in conservation.
U.S. Department of Defense (DoD) Readiness and Environmental Protection Integration Environmental and military considerations near DoD facilities	Projects are located around military lands and include designated Sentinel Landscapes, a nationwide federal, local, and private collaboration dedicated to promoting natural resource sustainability and working lands in areas surrounding military installations.	DoD Readiness and Environmental Protection Integration program coordinates mutually beneficial programs and strategies to preserve, enhance, or protect habitat and working lands near military installations, and reduce, prevent, or eliminate restrictions that inhibit military testing and training.

[a] U.S. Department of Interior programs are indicated by asterisks.
[b] Since this report entered review, the committee has learned that the Southeast Climate Consortium is no longer currently funded, though they will have an opportunity to apply for future funds again. Therefore the RISA program now currently supports 10 regional research teams, not 11. Because the committee learned of this after the report entered review, references to the Southeast Climate Consortium remain throughout this chapter.

BOX 5.1
The Five Attributes of Programs Considered

To gauge the similarity or overlap among other programs and the LCCs, the committee primarily considered the following five attributes:

1. **Extent of land area covered**
   The LCCs are charged with "creating a national network" and cover the entire U.S. land area as well as parts of Canada and Mexico. Some other programs are designed to operate in smaller, more specific geographic areas.

2. **Emphasis on scientific research**
   The LCCs were created, at least in part, to "develop and provide integrated science-based information" and "facilitate the exchange of applied science in the implementation of conservation strategies and products." Hence, the committee evaluated the emphasis on scientific research, as judged from the frequency with which research is mentioned in the goals and mission, and in competitively awarded funding (if applicable).

3. **Emphasis on climate as a driving issue**
   Secretarial Order 3289 launched the LCCs to address climate change and other landscape-scale issues and stressors.

4. **Emphasis on conservation within stated priorities and demonstrable activities**
   While some agencies have a mandate to manage for multiple uses or for specific governmental mission(s), LCCs are focused on the sustainability of our economy, land, water, wildlife, and cultural resources, using science to deliver conservation strategies.

5. **Degree to which the program's governance is concentrated in a single agency**
   In some programs, governance and management are concentrated in a single agency (e.g., USDA Regional Climate Hubs) versus having a decision-making body representing a wide range of organizations (e.g., LCC steering committees). LCCs are directed to facilitate collaboration among resource management agencies and organizations. They are to develop shared, landscape-level conservation objectives.

TABLE 5.2 Description of the Five Programs Relative to the Five Attributes

	Program				
Attribute	Landscape Conservation Cooperatives (LCCs)	FWS Migratory Bird Joint Ventures (JVs)	USGS Climate Science Centers (CSCs)	NOAA Regional Integrated Sciences and Assessments (RISAs)	FWS Fish Habitat Partnerships (FHPs)
Extent of land area covered	Cover all U.S. states and territories, and parts of Canada and Mexico	Covers essentially all of the United States, Canada, and a large part of Mexico	Cover all U.S. states and territories	Cover ~75% of U.S. land area	Nominally all of the United States but in reality, mostly riparian areas and adjacent lands
Emphasis on conducting scientific research	Fund extramural research and other activities	Some conduct original research, others fund extramural research and other activities	Fund extramural research and other activities; more research than LCCs	Conduct original research	No research
Emphasis on climate as a driving issue	Varies across the network as determined by each LCC's stakeholder group	Low priority	Prominent	Prominent	Low priority
Emphasis on conservation within stated priorities and demonstrable activities	Conservation is a top priority	Conservation is a top priority	Conduct research, in part to support conservation, but do not do conservation projects	Little to no emphasis on conservation	Conservation is a top priority
Degree to which the program's governance is concentrated in a single agency	LCCs are each steered by a broadly drawn stakeholder group	JVs are each steered by a broadly drawn stakeholder group	USGS–university cooperative agreement; USGS side has broadly drawn stakeholder advisory council for each CSC	Housed at universities	FHPs are each steered by a broadly drawn stakeholder group

NOTE: FWS = U.S. Fish and Wildlife Service; NOAA = National Oceanic and Atmospheric Administration; USGS = U.S. Geological Survey.

Conservation Act[2] in 1988 in support of the NAWMP and to provide grants to carry out wetlands conservation projects. Mexico officially joined the NAWMP in 1994. Although the plan is international in scope, planning and implementation occur at regional levels.[3] These regional efforts, known as Joint Ventures, seek to bring together the relevant agencies, organizations, and stakeholders to conserve habitat for migratory birds, other wildlife, and people within their region.[4]

The Joint Ventures enable partners to engage, both together and independently, in activities that support specific bird and bird habitat conservation goals within their geographic region. The activities conducted by the Joint Ventures and their partners range from biological planning, conservation design, habitat conservation (i.e., implementation), communication and outreach, monitoring, evaluation, and applied research.[5]

The first Joint Venture was formed in 1986, and today there are 22 habitat-based Joint Ventures (see Figure 5.1); 18 in the United States and 4 in Canada.[6] Each Joint Venture is governed by a management board that directs its activities and oversees its implementation planning; membership of the boards is determined within each Joint Venture and includes representatives from the organizations participating in the Joint Venture partnership. Joint Ventures also have one or more technical committees that provide scientific advice related to conservation goals. Each Joint Venture is staffed by a coordinator, and many also include additional positions in science coordination, science delivery, or communication.[7]

An Association of Joint Venture Management Boards has been organized that includes management board chairs and members from each of the Joint Ventures.[8] This association develops common messages about the impacts and successes of Joint Ventures and provides a forum to share les-

[2] 16 U.S.C. §§ 4401-4414.

[3] See http://www.fws.gov/birdhabitat/NAWMP/index.shtm.

[4] See http://mbjv.org/who-we-are.

[5] See http://www.fws.gov/policy/721fw6.html.

[6] See http://mbjv.org/who-we-are.

[7] See http://mbjv.org/who-we-are; http://www.fws.gov/policy/721fw6.html.

[8] See http://mbjv.org/who-we-are/networks.

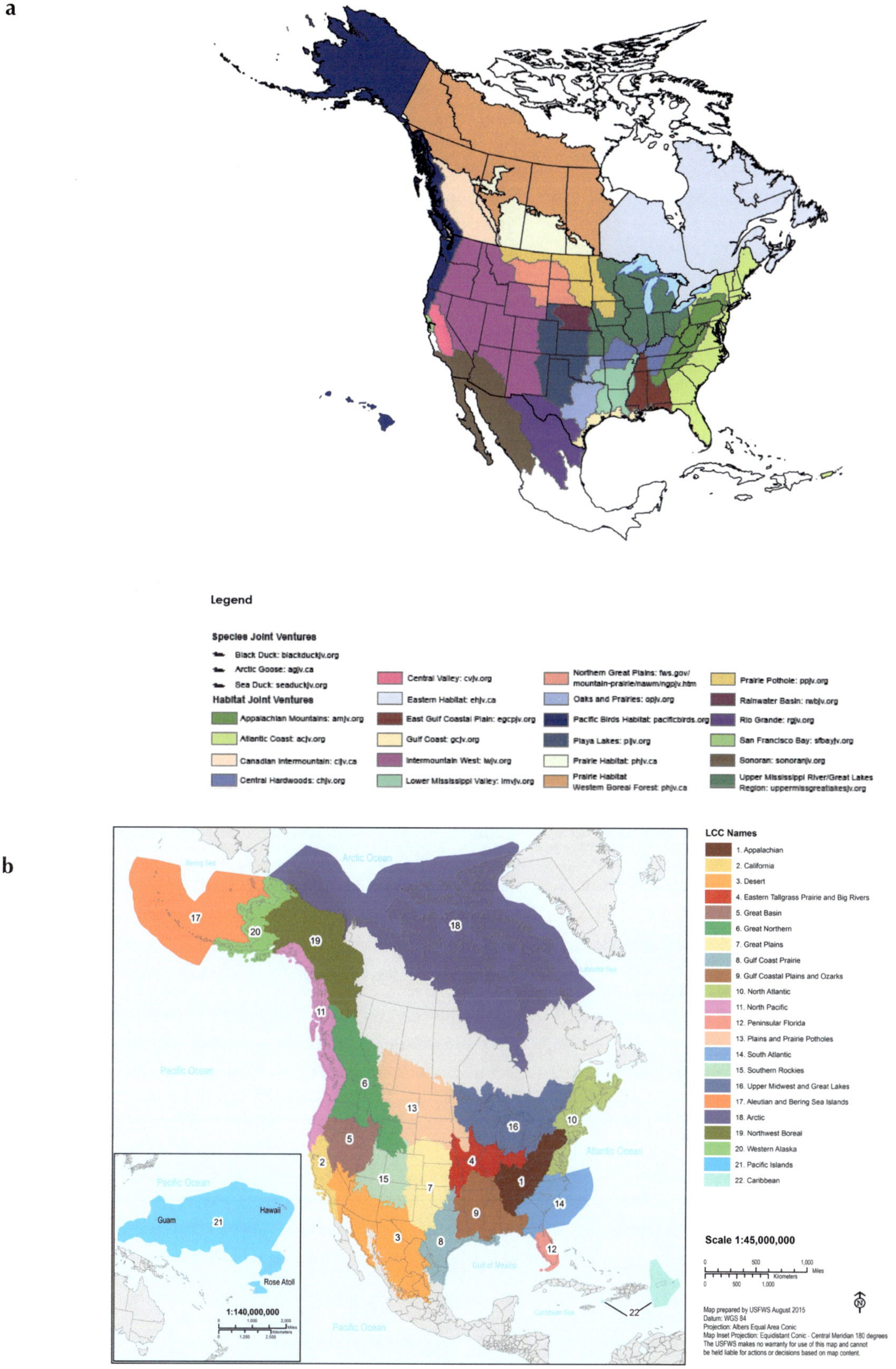

FIGURE 5.1 (a) Map of the Joint Ventures and (b) map of the Landscape Conservation Cooperatives shown for comparison.
SOURCES: http://www.fws.gov/birdhabitat/JointVentures/files/JointVentureFactSheet.pdf; https://www.sciencebase.gov/catalog/item/55b943ade4b09a3b01b65d78.

sons learned across individual Joint Ventures, but it does not direct their efforts.[9] Joint Ventures receive support through U.S. congressional appropriations to the FWS,[10] as well as additional federal, state, and private funding provided by partner organizations. The FWS typically provides funding for the coordinator and basic program infrastructure.[11]

Similarities with LCCs: In the 2012 revisions to the NAWMP, the Joint Venture approach of advancing conservation objectives through regional partnerships was cited as serving as a model for the LCCs and other emerging conservation efforts (NAWMP, 2012). Indeed, through communication with FWS personnel and a range of stakeholders, the committee learned that the Joint Ventures served as an inspiration for the LCC program. As a result, the LCCs naturally resemble the Joint Ventures in many ways. The Joint Ventures, like the LCCs, are self-directed partnerships organized at a regional level that together cover the entire United States and extend into Canada and Mexico. The governance structure for each program is similar: decisions for individual Joint Ventures and LCCs are made by representatives from the organizations participating in each partnership (management boards for the Joint Ventures and steering committees for the LCCs). Some Joint Ventures engage both a steering committee and a management board, the latter playing the role of implementing joint priorities (see Chapter 6 for a discussion of the Atlantic Coast Joint Venture). This structure is designed to ensure that the focus and specific project objectives for each Joint Venture and LCC address regional needs rather than being driven by national-level priorities. While each program includes some national-level organization (the Association of Joint Venture Management Boards and LCC Council), they are specifically designed to support efforts of the regional partnerships by enabling sharing of lessons learned, increasing awareness of the Joint Venture or LCC programs and related successes, and encouraging support of and participation in regional efforts. Many individual Joint Ventures and LCCs also have a science or technical committee to provide guidance on the best available science to inform partnership decisions.

The LCCs are described in LCC Information Bulletin No. 1 (Office of the Science Advisor, 2010) as "applied conservation science partnerships" designed to support conservation at the landscape scale. Similar to the Joint Ventures, they are intended to facilitate collaboration among the DOI bureaus and with other resource management agencies and organizations at the federal, state, and local levels, creating partnerships that are capable of "accomplish[ing] conservation objectives that no single LCC, nor any agency or organization, could accomplish alone." Also similar to the LCCs, the Joint Ventures operate across national borders to coordinate conservation with Canada and Mexico.

Differences from LCCs: Although the LCCs have much in common with the Joint Ventures, the primary difference between these programs is in their programmatic scope. The Joint Ventures have a relatively narrow focus on migratory bird and bird habitat conservation.[12] In contrast, the LCCs have a broader focus on natural and cultural resources more generally.[13] While individual LCCs may engage in some projects that support conservation of birds or bird habitats, their purview is much broader and may focus on other conservation issues deemed to be a priority in a particular area; indeed a review of LCC projects demonstrates a broad range of supported topics.[14] Joint Ventures have also been actively implementing conservation and monitoring activities, distinguishing them from the LCCs, which have so far focused more on collaboration, information sharing, and science development (see additional discussion in Chapter 6). Although LCCs intend to catalyze conservation, they are not yet viewed as a program that implements conservation the same way the Joint Ventures are perceived.

Coordination with LCCs: The committee found several existing avenues for limited coordination between the LCCs and Joint Ventures. An examination of the rosters of LCC advisory committees reveals many names with Joint Venture affiliations, either staff or management board members. In addition, some LCCs have funded joint projects with Joint Ventures: the list of LCC-funded projects includes 23 projects that mention Joint Ventures as partners or joint funders. Most of those (12) are with the California LCC.

Fish Habitat Partnerships

In 2006, a coalition of state, tribal, territorial, and federal government representatives as well as anglers, conservation groups, and scientists developed a National Fish Habitat Action Plan. The purpose of the plan was to encourage "voluntary, non-regulatory, science-based action to protect, restore, and enhance America's aquatic systems."[15] The effort was spearheaded by states through the Association of Fish and Wildlife Agencies (AFWA) in cooperation with the FWS and NOAA, which served as the primary liaisons to other federal agencies and other partners.[16]

---

[9] See http://mbjv.org/who-we-are/networks.

[10] Congressionally appropriated funds for both the LCCs and Joint Ventures are provided through an appropriation for "Resource Management." In a congressional report that accompanies the appropriating legislation, Congress clarifies how those funds are to be divided among various programs, including the LCCs and Joint Ventures (e.g., H.R. 2822, 114th Cong. 2015 and H.R. Rep. No. 114-170, p. 13).

[11] See http://www.fws.gov/policy/721fw6.html.

[12] See http://www.fws.gov/birdhabitat/JointVentures/index.shtm.

[13] See http://lccnetwork.org/about.

[14] See http://lccnetwork.org/projects.

[15] See http://fishhabitat.org/sites/default/files/www/NFHP_AP_Final_0.pdf.

[16] See http://fishhabitat.org/sites/default/files/www/National%20Fish%20Habitat%20Action%20Plan%20-%20National_Fish_Habitat_Action_Plan.pdf.

The plan called for the organization of a network of regional FHPs focused on important aquatic habitats and species. These partnerships serve as the working units of the National Fish Habitat Action Plan and are overseen by the National Fish Habitat Board. The National Fish Habitat Board comprises 22 members representing a variety of stakeholder groups, including state and federal agencies, tribal governments, conservation groups, resource managers, and academia. Members of the board are approved by an executive leadership team that includes the president and executive director of AFWA, the assistant administrator for Fisheries at NOAA, and the director of the FWS. The purpose of the National Fish Habitat Board is to provide leadership and coordination, approve and support FHPs, establish interim and long-term national conservation goals, support regional goals, mobilize support, and measure and communicate the status and needs of fish habitat. Staff support for the National Fish Habitat Board is provided by AFWA, the Michigan Department of Natural Resources, the FWS, NOAA's National Marine Fisheries Service, and USGS.[17]

By 2012, when an updated second edition of the National Fish Habitat Action Plan was released, the National Fish Habitat Board had approved 18 regional FHPs, and in 2014, a 19th regional FHP in Southeast Alaska was approved. Among the 19 FHPs, all 50 states are represented in at least one partnership.[18] Of the 19 FHPs, 18 are based either on a particular geographic region or a particular species of fish. The Reservoir Fish Habitat Partnership is considered to be a "system-based" partnership that includes reservoirs throughout the country (see Figure 5.2).

Similarities with LCCs: Like the LCCs, the National Fish Habitat Partnership was modeled after the Joint Ventures program,[19] and therefore, shares many traits with both the Joint Ventures and the LCCs in terms of structure and function. Like the LCCs and the Joint Ventures, the individual Fish Habitat Partnerships are considered self-directed. The updated National Fish Habitat Action Plan (2012) describes them as "self-identified, self-organized, and self-directed communities of interest formed around geographic areas, keystone species, or system types." They also resemble the Joint Ventures and LCCs in that they are overseen by a board at the national level, but they are primarily intended to operate at a regional scale through individual, cross-jurisdictional partnerships of diverse stakeholders. The individual FHPs are also typically led by a steering committee.

Differences from LCCs: As with the Joint Ventures, the primary difference between the FHPs and LCCs is their scope. Because these partnerships focus on a particular taxonomic group, their focus is also not as broad as the LCCs, which focus on natural and cultural resources generally. Whereas several LCCs include partners in Canada or Mexico, with one exception (the Great Lakes Basin Fish Habitat Partnership) the FHPs focus solely on U.S. waters. Another difference from the LCCs is that there is substantial geographic overlap among the FHPs. In contrast to the LCCs and according to the National Fish Habitat Action Plan (2012), the primary focus of the FHPs is to implement conservation projects.

Coordination with LCCs: Despite the differences in geography and focus just noted, the confluence of conservation priorities sometimes brings LCCs and FHPs to work together. The committee found multiple references to collaborations between LCCs and FHPs, including but not limited to coordination on fish habitat assessments. For example, the Plains and Prairie Potholes LCC worked with FHPs to develop advanced fish habitat assessment models to support efficient, targeted, and strategic management of aquatic resources.[20] The committee also noted some overlap in the membership of the Fish Habitat Partnership steering committees and the membership of steering committees of the LCCs.

Climate Science Centers and National Climate Change and Wildlife Science Center

USGS's National Climate Change and Wildlife Science Center (NCCWSC) serves as the managing entity for eight regional CSCs. The CSCs were established by the same Secretarial Order (No. 3289; see Appendix E) that established the LCCs, and were intended to provide information about climate change to all DOI bureaus. The LCCs and CSCs were intended to be complementary, with the CSCs focused more on research and LCCs on delivering information and convening partners for developing shared conservation strategies. Like the LCCs, the eight CSCs together span the entire contiguous United States from coast to coast, as well as Alaska, Hawaii, and the U.S. territories (see Figure 5.3). The eight CSCs are all housed at research universities that were selected through a competitive process.

Each CSC has the responsibility and opportunity to establish its own goals and research priorities in consultation with its Executive Stakeholder Advisory Committee, which entirely comprises federal, state, and tribal representatives. LCC coordinators are key stakeholders on the Executive Stakeholder Advisory Committees and as such serve an important function in ensuring that each CSC is responsive to LCC needs. NCCWSC has developed its own strategic plan, which includes some goals and aspirations for the CSCs. It also has its own advisory committee, but unlike the regional

[17] See http://fishhabitat.org/sites/default/files/www/NFHP_AP_Final.pdf.

[18] See http://fishhabitat.org/sites/default/files/www/NFHP_AP_Final.pdf.

[19] See http://www.fws.gov/fisheries/whatwedo/NFHAP/nfhap_who.html.

[20] See http://www.fws.gov/midwest/lccpartnerships.htm.

a

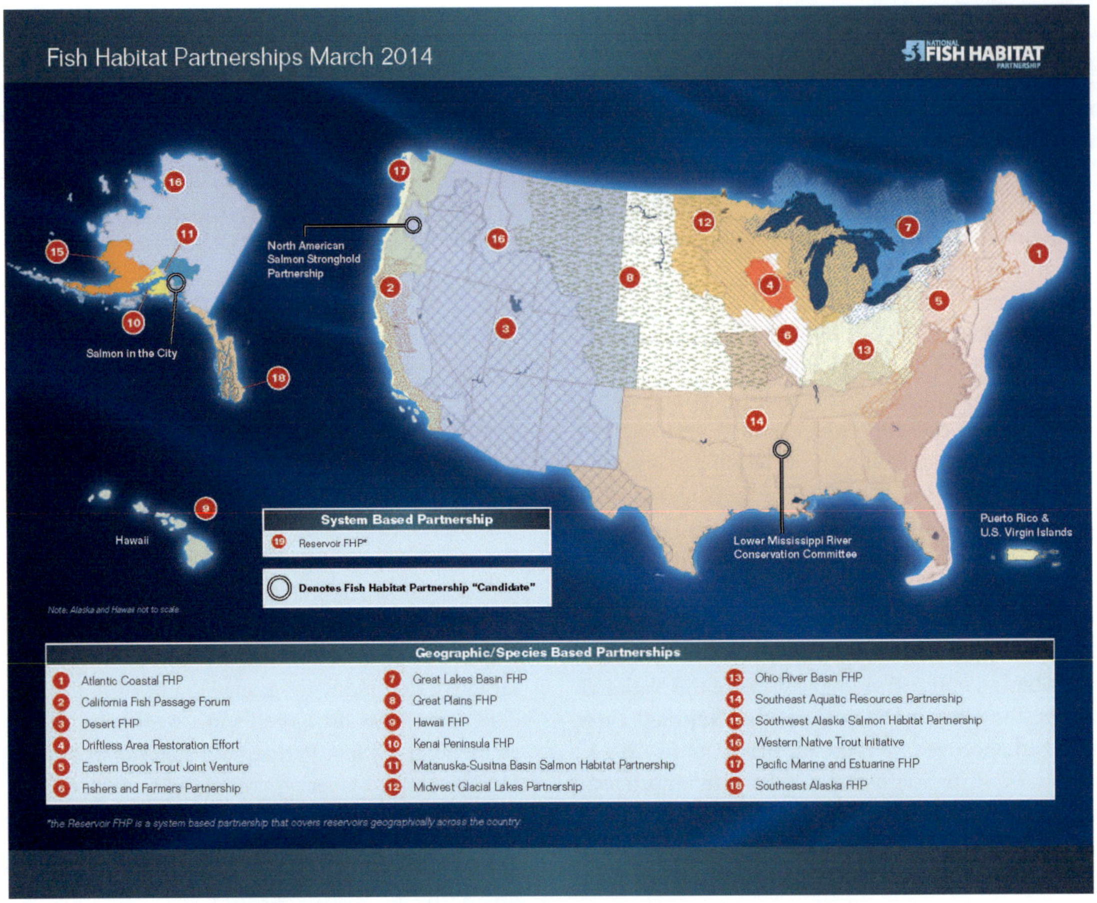

b

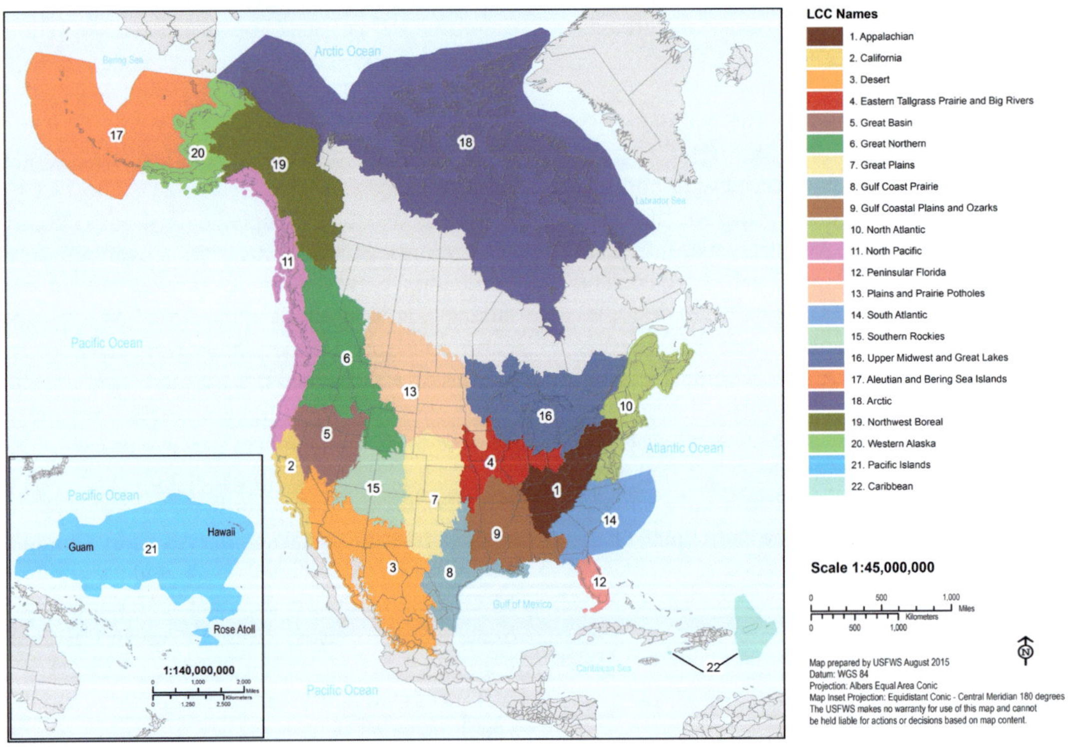

FIGURE 5.2 (a) Map of the Fish Habitat Partnerships and (b) map of the Landscape Conservation Cooperatives shown for comparison. SOURCES: http://fishhabitat.org/sites/default/files/FHP_Map_Main_14_1.pdf; https://www.sciencebase.gov/catalog/item/55b943ade4b09a3b01b65d78.

a

b

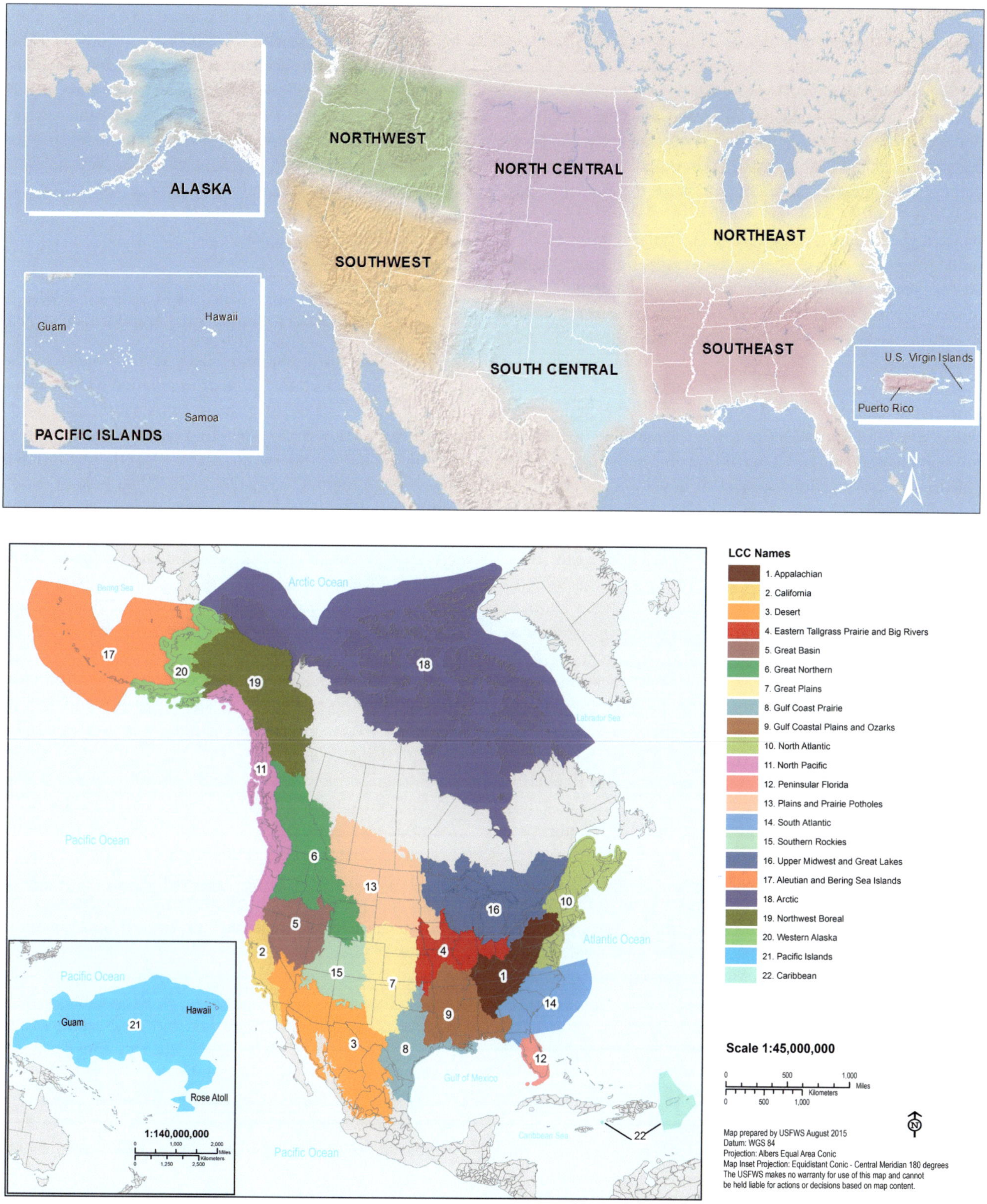

FIGURE 5.3 (a) Map of the Climate Science Centers and (b) map of the Landscape Conservation Cooperatives shown for comparison. SOURCES: https://edit.doi.gov/csc/centers; https://www.sciencebase.gov/catalog/item/55b943ade4b09a3b01b65d78.

Executive Stakeholder Advisory Committees, none of the members are leaders of LCCs or its national program.[21]

Similarities with LCCs: Both the LCCs and the CSCs were created by the same Secretarial Order and cover the geography of the United States and its territories; both are housed in DOI; both award external grants; and both are guided by steering committees that have some ability to make recommendations including funding decisions, with the final decisions resting on the regional director (for the FWS). Furthermore, members of the respective steering committees comprise similar stakeholder groups; thus, potentially resulting in the identification of similar priorities. Both have an emphasis on climate science to support decisions, including cultural and natural resource management.

Differences from LCCs: As described in the Secretarial Order, the 22 LCCs and 8 CSCs were intended to be distinct but complementary. LCCs were intended to focus more on convening conservation partners to develop common conservation priorities and Landscape Conservation Designs as well as the applied science and tools to inform these conservation priorities. In contrast, CSCs were intended to focus on carrying out the research needed to support conservation and resource management in the face of climate change. CSCs are not designed to develop conservation strategies and Landscape Conservation Designs, but focus primarily on research. Structurally, CSCs consist of a university component competitively awarded on a 5-year cooperative agreement, and of a small number of USGS staff who direct USGS funding to a portfolio of research projects. By contrast, LCCs have no university component. The Executive Stakeholder Advisory Committees for CSCs consists solely of state, federal, and tribal representatives; the steering committee of LCCs includes much broader participation (e.g., nongovernmental organizations [NGOs] and private sector). Whereas all CSCs have a significant focus on climate, there is wide variety among LCCs in the prominence of climate issues in their communication, activities, and funded projects. A comparison of word clouds created from the mission statements of the individual LCCs and CSCs (after removing geographically specific terms and every occurrence of the name of the entity) reveals a much greater emphasis on climate, tools, and science in the CSCs, and much greater emphasis on conservation in the LCCs, reflecting the needs of their stakeholders (see Figure 5.4).

Coordination with LCCs: Relevant LCC coordinators are usually on the steering committee for CSCs, though not usually the reverse (since some CSC regions touch as many as seven LCCs). In some instances, CSCs and LCCs have jointly funded projects. The list of LCC-funded projects includes seven that were jointly funded with CSCs, mostly in Alaska. An additional two LCCs listed CSCs as leverage, and a number of LCC projects list CSCs as participants. At least 10 persons listed on the rosters of LCC advisory councils are directors or other leaders of CSCs. The LCCs and CSCs have also coordinated on meetings, workshops, and symposia. For example, the Southwest Climate Summit was organized jointly by the Southwest CSC, California Nevada Climate Applications Program, and the California LCC. The Pacific Island LCC and Pacific Island CSC have also cohosted two Climate Science Symposia.

[21] See https://nccwsc.usgs.gov/content/acccnrs-member-list.

National Oceanic and Atmospheric Administration's Regional Integrated Sciences and Assessments Program

The RISA program dates back to 1995 and was developed within NOAA to "pioneer innovative mechanisms for enhancing the value of climate information and products for understanding and responding to these challenges at the regional scale" (Binder et al., 2009). The RISA program now supports 11 regional research teams[22] that help expand and build the nation's capacity to prepare for and adapt to climate variability and change. RISA teams work with public and private user communities to develop knowledge on impacts, vulnerabilities, and response options through interdisciplinary research and participatory processes. Climate information can inform decisions intended to promote adaptation to a changing environment, but only if the climate research community and decision makers work together to understand each other's needs and limitations. Successful outcomes from the RISA program are due in part to their ability to create lasting relationships with decision makers from the public and private sectors including local, regional, and state governments, federal agencies, tribal governments, utilities, the business community, and national and international nonprofit organizations (Pulwarty et al., 2009; Meadow et al., 2015). Through these relationships, RISA teams learn about specific decision contexts within and across different sectors of society, advancing the overall understanding of the use of science.[23]

Similarities with LCCs: Like LCCs, RISAs aim to serve a variety of functions: supporting and conducting research, developing tools or other approaches to support decision makers, and convening stakeholders, to name a few. Each RISA has the latitude to set its own priorities, choose which stakeholders to engage, and to some extent create its own approach to governance. Similarly, LCCs also have the

[22] Since this report entered review, the committee has learned that the Southeast Climate Consortium is no longer currently funded, though they will have an opportunity to apply for future funds again. Therefore the RISA program now currently supports 10 regional research teams, not 11. Because the committee learned of this after the report entered review, references to the Southeast Climate Consortium remain throughout this chapter.

[23] See http://cpo.noaa.gov/ClimatePrograms/ClimateandSocietalInteractions/RISAProgram/AboutRISA.aspx.

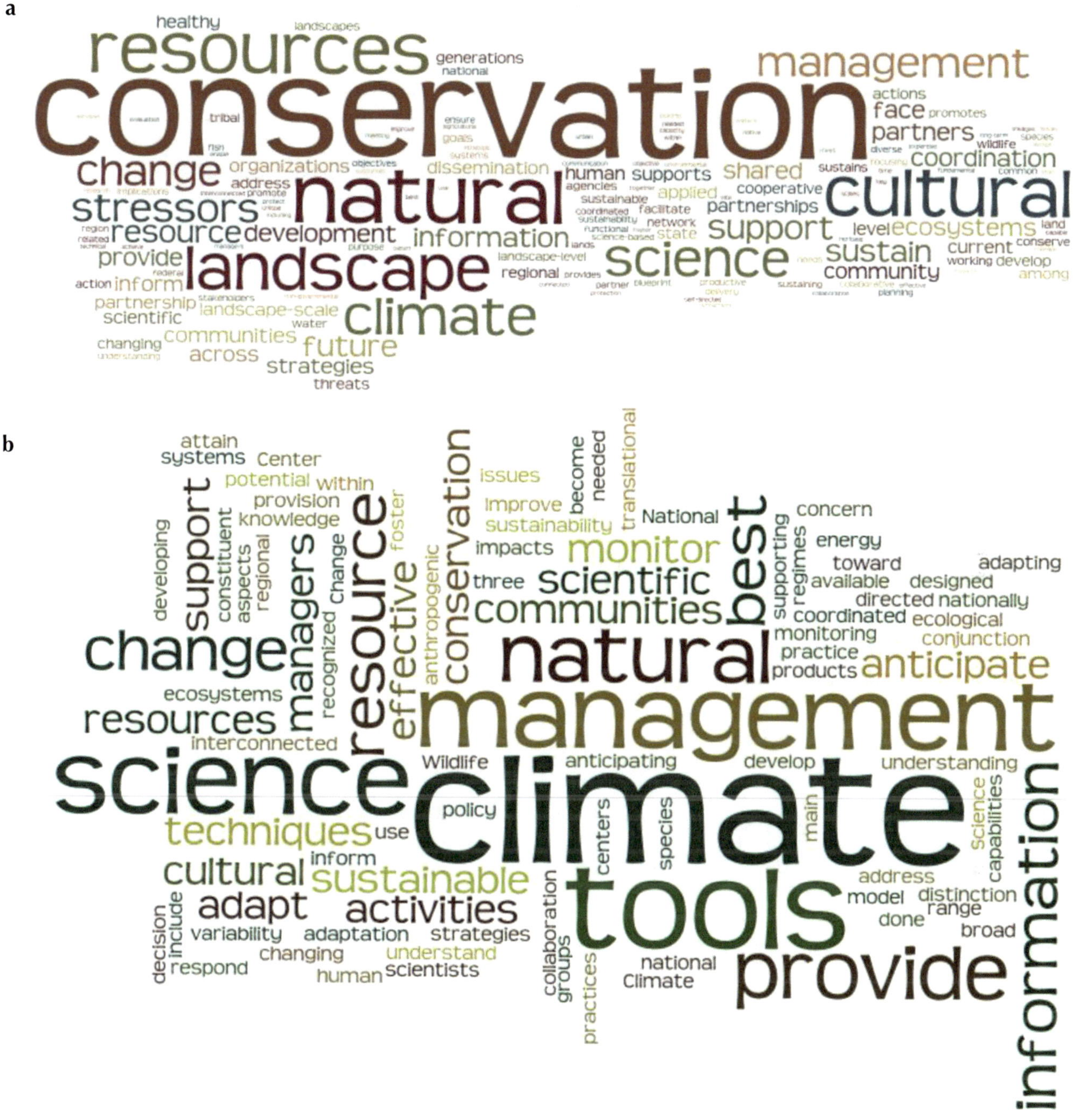

FIGURE 5.4 Word clouds derived from the mission statements of (a) the 22 Landscape Conservation Cooperatives and (b) the 8 Climate Science Centers.

authority as self-directed partnerships to develop their own diverse priorities and, to a lesser extent, however, to determine their own leadership.

Differences from LCCs: The RISA program is quite different from LCCs in mission, governance, topical focus, and goals. RISAs, whose core participants are based at universities, are much more focused on research, and are created through a competitive process by proposing teams of university scientists, not by an agency. RISAs, unlike CSCs and LCCs, do not cover the entire country, as can be seen in Figure 5.5. The teams of university scientists propose regional boundaries, select the key participants, determine the composition and structure of their advisory councils (if any), and determine the priorities. RISAs do not themselves provide external funding for projects the way LCCs do; instead, they carry out their research agenda and other activities with the proposing team and often with leveraging (as with LCCs). Broadly, RISAs focus more directly on human considerations like urban water availability, public health, and community resilience; and RISA teams generally include social scientists.

Coordination with LCCs:

RISAs and LCCs primarily coordinate by convening joint meetings, workshops, conferences, and training events. For example:

- The Great Basin LCC and California-Nevada Assessment Program have organized a series of six Great Basin Climate Forums and two Tribal Climate Adaptation Workshops; combining forces has led to more effective dissemination and application of climate information, participants assert.
- The Desert LCC, and the Southwest and South Central RISAs co-convened a U.S.-Mexico binational climate change adaptation workshop, concentrating on the Big-Bend stretch of the Rio Grande Basin.[24]
- LCC coordinators serve on steering committees of at least two RISAs: the Alaska RISA[25] and the Northwest Climate Impacts Research Consortium,[26] and conversely, RISA participants serve on science advisory committees for at least two LCCs (the Northwest Climate Impacts Research Consortium on the North Pacific LCC, and Western Water Assessment on the Southern Rockies LCC[27]).

COORDINATION AMONG REGIONAL PROGRAMS

In this section, we examine in more detail the coordination among LCCs, CSCs, and RISAs in two geographic regions. We selected the Pacific Islands because the geographic boundaries of the Pacific Islands LCC, the Pacific Islands CSC, and the Pacific RISA are nearly identical, affording an organizationally simple framework to examine the interactions among the three types of regional programs. Because the alignment of geographic boundaries for these programs is not typical, we also analyzed in more depth the coordination among the Southeast CSC and its six affiliated LCCs and three RISAs.

Pacific Islands

The geographic extent includes the state of Hawaii and U.S.-affiliated Pacific territories, including Guam, American Samoa, and the Federated States of Micronesia. The committee learned from conversations with stakeholders involved in these programs that people in this vast, far-flung, and sparsely populated region rely on and value remote connections, personal networks, and collaborations. Therefore, organizations offering such networking, as well as decision-relevant research and products, like the Pacific Islands LCC, Pacific RISA, and Pacific Islands CSC, are valued.

Several examples serve to illustrate the collaboration and synergies of these three organizations. The Pacific Islands LCC and Pacific Islands CSC have cohosted two Climate Science Symposia.[28] The Pacific Islands LCC and Pacific RISA collaborated on the Pacific Islands Regional Climate Assessment,[29] which was the regional contribution to the third U.S. National Climate Assessment. The Pacific Islands Regional Climate Assessment was a collaborative effort engaging federal, state, and local government agencies, NGOs, academia, businesses, and community groups to inform and prioritize their activities in the face of a changing climate (Keener et al., 2012).Three technical workshops were held between November 2011 and January 2012, and the Regional Climate Assessment was released later in 2012. Most of the funding came from NOAA, through the Pacific RISA, and the Pacific Islands LCC was also a financial contributor. Because the Pacific Islands CSC was in its infancy at the time, it had a small role in the Regional Climate Assessment.

The research emphases of each entity are somewhat distinct and evolving. The Pacific Islands region's LCC, CSC, and RISA all share an interest in the physical drivers of change, including sea level rise, ocean wave characteristics, and climate downscaling. However, the methods, applications, and partners of the three organizations are justifiably distinct, and to a large degree complementary.

Current Emphases of the Pacific RISA: Research emphases include

1. Regional climate modeling. Techniques using Global Climate Models (GCMs) to generate future climate projections may be of less value over small islands, which are too small to be represented as land in the GCMs. The Pacific RISA supports the use of the Hawaii Regional Climate Model (HRCM) to generate high-resolution climate data for the Hawaiian Islands, Guam, and American Samoa at spatial resolutions as fine as 1 km, a spatial scale necessary to take into account steep topography and diverse microclimates.
2. Future groundwater recharge. Using HRCM projections, Pacific RISA partner researchers at the USGS Pacific Islands Water Science Center and the University of Hawaii Water Resources Research Center used calibrated soil–water balance models for the island of Maui to calculate ground-

[24] See http://cpo.noaa.gov/Partnerships/International/TheNorthAmericanClimateServicesPartnership%28NACSP%29_b/RioGrande-RioBravoRegionalPilotArea.aspx; http://www.climas.arizona.edu/blog/notes-field-preparing-climate-change-along-us-mexico-border.

[25] See https://accap.uaf.edu/about.

[26] See http://pnwcirc.org.

[27] See http://southernrockieslcc.org/about-srlcc/science-workgroup.

[28] See http://apdrc.soest.hawaii.edu/PICSC/symposium.php; http://apdrc.soest.hawaii.edu/PICSC/news/Review_Symposium.pdf.

[29] See http://www.pacificrisa.org/projects/pirca.

a

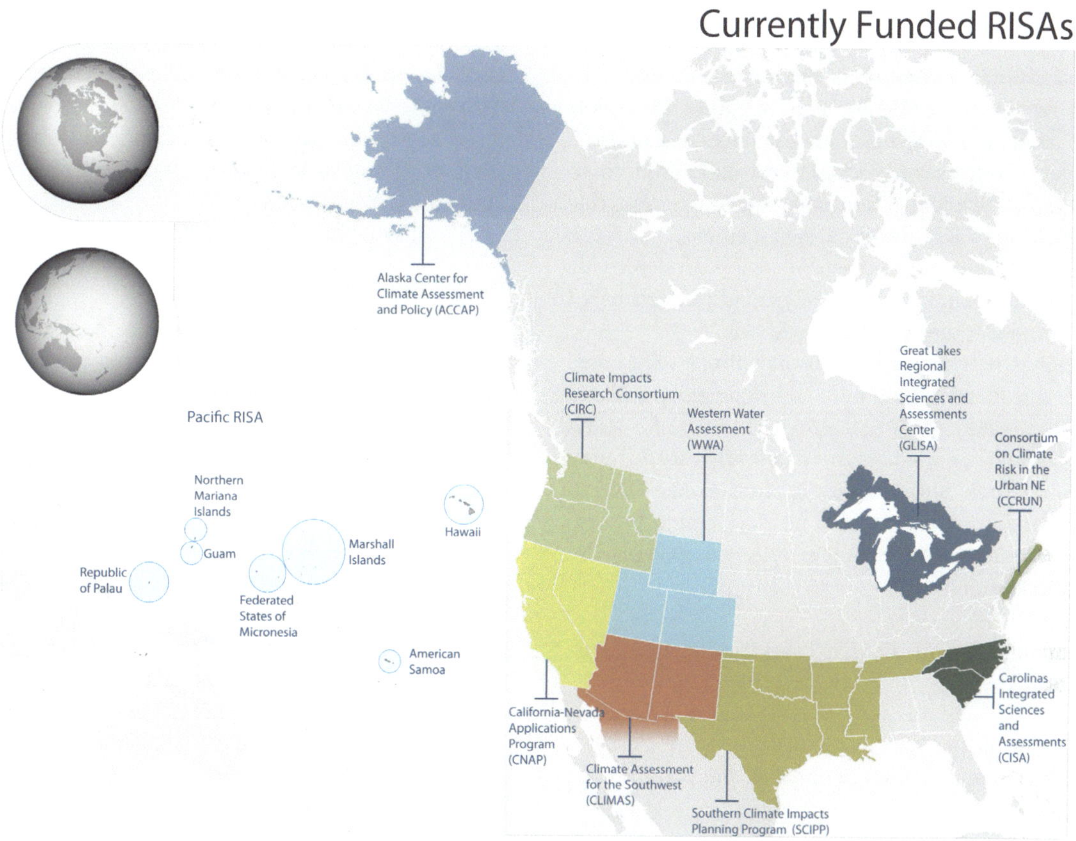

b

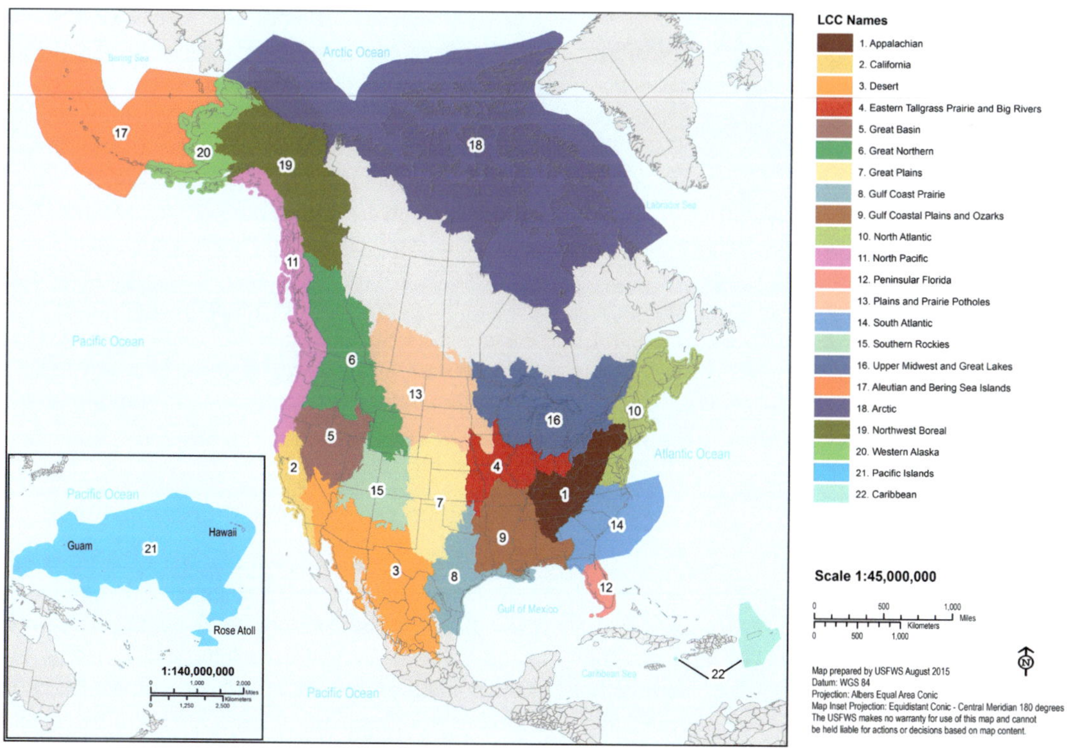

FIGURE 5.5 (a) Map of the 11 Regional Integrated Sciences and Assessments teams and (b) map of the Landscape Conservation Cooperatives shown for comparison.
SOURCES: http://cpo.noaa.gov/ClimatePrograms/ClimateandSocietalInteractions/RISAProgram/RISATeams.aspx; https://www.sciencebase.gov/catalog/item/55b943ade4b09a3b01b65d78.

water recharge under future land and water management scenarios.

3. Land-use and hydrology scenario development. To ensure that results from climate and hydrological models address the needs of Maui and state-level decision makers, Pacific RISA researchers worked with stakeholders to generate a set of feasible future Maui land-use scenarios relevant to groundwater resource management. Land-use maps represent future management decisions and provide the spatial environment across which groundwater recharge is calculated under future climate conditions.

4. Regional network maps. To map flows of climate information and identify key hubs and potentially isolated groups in the greater Pacific Islands region, Pacific RISA researchers tracked communications patterns across different sectors and countries. Survey analysis of more than 300 climate change professionals revealed network connectedness and perceived community resiliency, climate change risk, and sense of community.

Current Emphases of the Pacific Islands LCC: Research emphases include

1. Mapping potential ranges of native species and invasive species under future temperature and precipitation projections;
2. Leading vulnerability assessments for rare species, native ecosystems, and keystone species;
3. Predicting future potential community composition within protected areas under different climate scenarios;
4. Identifying potential corridors linking present and future habitat;
5. Recommending conservation and acquisition priorities based on future climate and sea level; and
6. Developing adaptation strategies to protect biodiversity and cultural heritage across the Pacific.

Current Emphases of the Pacific Islands CSC: Research emphases include

1. High-resolution projections of future climate, sea level, and shoreline/inundation;
2. Estimates of low flow in ungaged streams;
3. Understanding and predicting vegetation change and management thereof; and
4. Coral reefs and other seascapes.

Consideration of Overlaps: As the above lists indicate, the Pacific Islands LCC, CSC, and RISA research is mostly complementary, with one exception: All three entities have funded projects on regional climate modeling for Hawaii and the Pacific Islands. However, from evidence available on the three programs' websites, this work was started by simulating climate in Maui with funding from the Pacific RISA because of a specific interest in island water issues. Funding from the Pacific Islands LCC and CSC allowed researchers to apply the modeling framework to other islands in Hawaii and beyond. The committee noticed in at least one case that the Pacific Islands LCC and CSC funded projects that appear to be nearly identical, giving the impression that the same work was funded twice and suggesting that coordination could still be improved.

Southeastern United States

In contrast to the Pacific Islands, the geographic boundaries of programs in the Southeast do not align. The geographic extent of the Southeast CSC (see Figure 5.6) includes all or parts of three RISAs (Carolinas RISA, Southeast Climate Consortium,[30] and Southern Climate Impacts Planning Program), and six LCCs (Appalachian, Gulf Coastal Plains and Ozarks, Gulf Coast Prairie, Peninsular Florida, South Atlantic, and Caribbean).

In the Southeast, the RISAs appear to be less involved with LCCs and the Southeast CSC than in the Pacific Islands. This should not be interpreted as a criticism; there appear to be strong and valid institutional and historical reasons for this separation. For instance, the Southeast Climate Consortium focuses on agriculture and on climate projections in the range of 1–12 months, a research focus and time horizon that the CSCs and LCCs are not undertaking.

Careful examination of the complete list of funded projects and activities by the Southeast CSC and the six LCCs revealed several projects with similarities. Specifically, seven projects provided variations of future climate projections, and four on the topic of sea level rise and associated saltwater habitat (marshes, mangroves). However, on closer examination, it became apparent that the seven climate-related projects are complementary. They produce different products over different geographic domains using, generally, different but application-appropriate techniques. As in the previous case study, the committee found that the Southeast CSC and one LCC funded similar projects. This also points to the need for the CSCs and LCCs to closely coordinate their external research-granting processes when the topics overlap.

Evaluation of Overlap and Coordination

The committee noted that the LCC program and other similar programs overlap in some areas. Below, the committee discusses the source of such overlaps; identifies benefits and costs that may be associated with overlaps, especially in the context of the LCCs; and determines that, with adequate coordination and rationale, the current extent of overlap is acceptable.

---

[30] Since this report entered review, the committee has learned that the Southeast Climate Consortium is no longer currently funded, though it will have an opportunity to apply for future funds again. Therefore, the geographic extent includes all or part of two currently funded RISAs, not three.

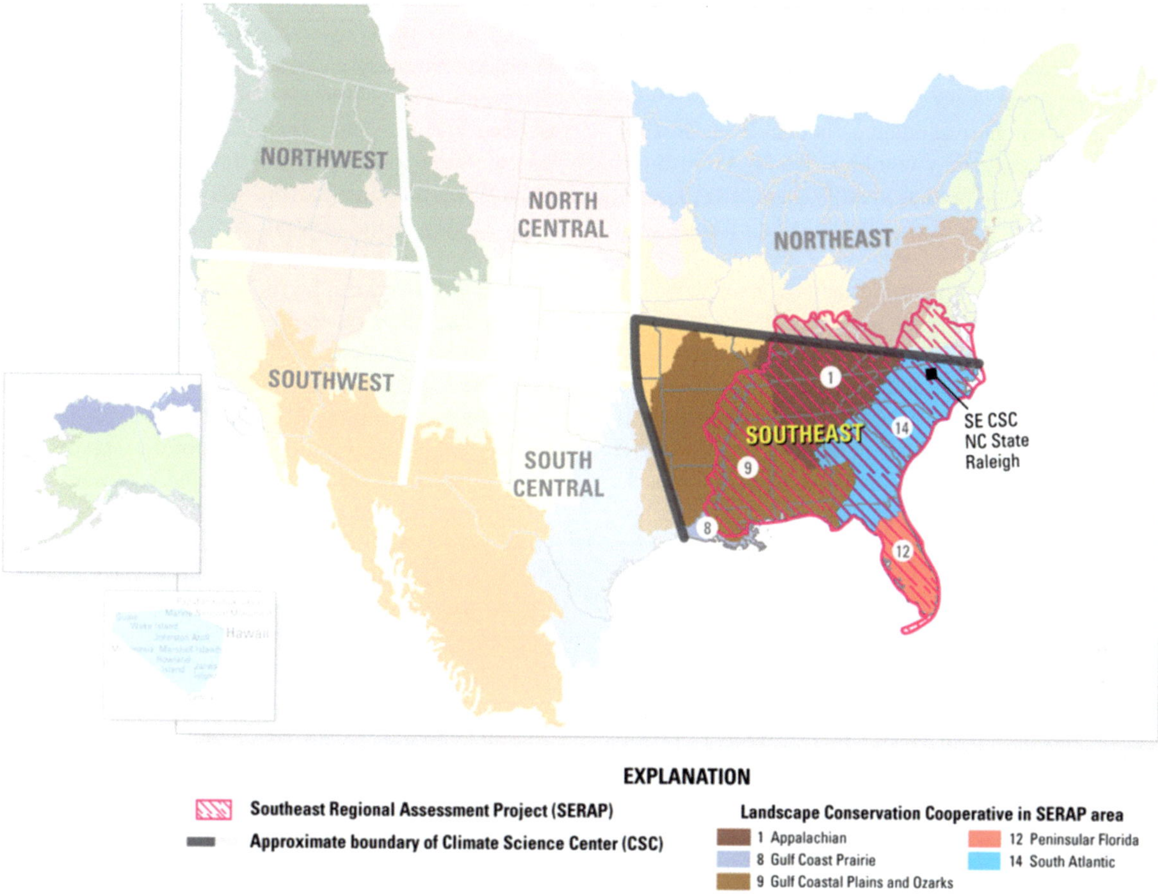

FIGURE 5.6 Map of the five LCCs, one CSC, and three RISAs in the Southeast.
SOURCE: https://globalchange.ncsu.edu/secsc/climate-science-centers.

Across the federal government, there are multiple directives to harness different authorities, programs, and funding sources with similar aims—ultimately, to promote conservation at a larger scale. Above, the committee illustrates a variety of such programs across federal agencies. In addition, there are numerous cross-agency efforts focused on either a geographic region (e.g., Chesapeake Bay or Gulf hypoxia; see Gulf hypoxia case study in Appendix B) or a species (e.g., sage-grouse; see case study in Appendix A). As discussed above, there are some similarities in projects, activities, and collaborations supported by LCC programs and other related federal programs.

By design, LCCs are collaborative conservation efforts (see Chapter 2) designed to identify strategic research and conservation priorities among partners, and help leverage funds. As a result, the research funded by LCCs is often collaborative with other agencies. Indeed, such collaborations were specifically encouraged in many LCC- and CSC-sponsored requests for proposals. Therefore, the committee was not surprised to find some overlap with other programs within the FWS, within other DOI bureaus, in other federal agencies, in state agencies, and in the nongovernmental conservation sector. Institutional redundancy and overlap are common in conservation, in natural resource management, and indeed in government organization more generally. The collaborative governance of the LCCs through each steering committee is designed to identify common goals and unmet research needs. The structure of the LCCs is intended to enhance communication and coordination of partners, and to streamline efforts in an increasingly complex field. Through this process, unnecessary overlap and redundancies are likely identified and avoided. Given the U.S. federalist structure, some degree of overlap between state and federal activities is both inevitable and desirable. Even within the federal system (and many state systems) there are significant institutional overlaps and shared regulatory (or management) spaces (Freeman and Rossi, 2012), as well as institutional gaps. In some cases, overlap is the product of federal law, and in this case, the Secretarial Order. Overlaps and redundancies can be beneficial or costly, depending on the context and societal goals.

There are administrative costs associated with redundancy; it is generally more efficient to have a single entity responsible for a mission or a set of objectives than to have multiple entities involved. It may also add confusion, lead to conflict among agencies, and complicate interactions with those seeking to engage in regulated actions. If the institutional landscape is especially complex, it may be difficult either to pinpoint authority or to identify gaps in regulatory coverage (Crowder et al., 2006). Fragmentation may also diffuse responsibility in ways that provide incentives for each of the multiple actors to underplay its own responsibility, avoiding difficult decisions in the hope that other actors will confront them (Buzbee, 2003). It makes planning difficult, because no one entity has a comprehensive view of the managed system. Fragmentation can make learning difficult if there are not well-established lines of communication. It can also make it hard to see, and especially address, the impacts of cumulative or synergistic threats. However, the landscape approach taken by the LCC Network aims to overcome such institutional fragmentation and the associated complexities and inefficiencies. The LCC Network Strategic Plan (LCC, 2014) specifically states that one tactic for creating high-functioning organizational culture for LCCs is to "[i]dentify institutional barriers and stovepipes that inhibit cross-agency collaboration and partnerships and seek to reduce, break-down, or overcome them."

The involvement of and overlap among multiple management agencies can also provide benefits; moreover, the negative aspects of redundancy just mentioned may pertain more to regulatory concerns than to the science-management and collaborative priorities of LCCs. As in systems engineering, managing redundancy might reduce the risk of catastrophic failure by allowing one actor to compensate for another's shortcomings (Landau, 1969). Moreover, institutional separation and overlap can encourage diversity of ideas and approaches, combating the tendency to fall into patterns of "group think," where assumptions go unexamined and viewpoints tend to converge on an unrealistic extreme (O'Connell, 2006). It can bring to bear the distinct expertise, cultures, and missions of multiple institutional actors. Decentralizing authority also reduces the risk of domination by one particular stakeholder group whose goals diverge from those of the broader policy community (Curtin, 2015). At the most basic level, it is more difficult and costly to influence several agencies than to influence one. In addition, fragmentation allows for differing agency cultures and missions, which in turn should facilitate the zealous pursuit of multiple goals. In a competitive funding environment, scientific or conservation proposals that meet the objectives of multiple organizations (e.g., multiple LCCs or an LCC and a CSC) and show leveraging potential may achieve higher rankings during review than those with narrower support.

The right institutional design question, then, is not whether there is redundancy or overlap, but whether the institutional system as a whole is able to effectively pursue the goals for which it was developed. The answer depends on whether the entities involved are able to communicate, coordinate, and collaborate effectively, not simply on whether there is some overlap or some tension in their missions. The discussion above illustrates generally positive attributes of overlap because there is evidence of communication and coordination to achieve multiple objectives. The steering committee meetings of the LCCs attempt to coordinate across many of the existing programs. Furthermore, workshops, conferences, and training events that are jointly organized and funded can be of higher quality and impact than any one organization could achieve alone. Through personal communications with invited speakers and conversations with stakeholders, the committee learned that multiple LCC participants view the LCCs as a forum for increased and more efficient communication among stakeholders who often come from multiple states and represent a variety of NGOs, federal and state agencies, tribal organizations, and other partnerships. Further, multiple participants described the LCCs as avenues for building relationships, fostering future partnerships, increasing trust, providing learning opportunities, and enhancing technical capacity.

However, coordination among the LCCs and other programs requires staff resources that some of the partners and stakeholders might lack. The committee learned from invited speakers as well as from conversations with stakeholders that a major challenge to evaluation of the LCC Network is the number of collaborative efforts already under way (e.g., by the Joint Ventures, Western Governors' Association, the Western Association of Fish and Wildlife Agencies, and AFWA). Despite the fact that many stakeholders recognize that the LCC Network fills an important need, some, in particular state agency representatives, expressed a concern that they do not have sufficient staff capacity to substantially contribute to collaborative efforts. This problem is exacerbated in states with multiple LCCs, and they find themselves having to prioritize among collaborative initiatives.

Moreover, some representatives from state agencies indicated the difficult trade-offs they face in terms of dedicating resources and staff time to directly undertaking projects or participating in one or more LCCs or other partnership efforts. For example, Alaska includes designations in five different LCCs; California, Oklahoma, Texas, and Wyoming each have four LCCs within their borders; and several other states have three. In light of multiple competing demands on capacity, particularly staff time and travel funds, some partners expressed that the cost of engaging in an LCC appear greater than the benefits. Also, because the LCCs are relatively new, many partners have not yet seen the direct benefit to their own priorities or responsibilities. For example, one stated that the scale at which LCCs aim to work (across large landscapes and long timescales) does not match well with state agency objectives, which by their nature must focus on smaller-scale, more immediate concerns. Others saw the LCCs as an added layer of bureaucracy when funds for science delivery could be administered through preexisting groups and processes.

As the LCCs mature and benefits can be realized and documented, it may become less challenging to justify the staff cost. Some states have already overcome the challenge of participating in multiple LCCs by appointing a single staff person as the designated liaison to all LCCs within a state. Enhancing the opportunities for partners to be actively engaged with the full suite of regional activities that intersect with their missions will be important to the longevity of these efforts. Potential actions could include holding joint meetings between neighboring LCCs or among LCCs and overlapping programs, increasing the capabilities for virtual meetings with real-time interactions, or adjusting boundaries if appropriate to achieving conservation objectives.

CONCLUSIONS

A wide variety of landscape-scale programs exist that are federally funded or initiated. In addition, national associations and regional partnerships have formed that also aim to scale up conservation activities. The diversity of efforts, partnerships, and approaches that the committee reviewed all serve a unique mission or set of goals regarding landscape and habitat conservation. Therefore, the committee concludes that despite some overlap, none of the programs are fully redundant. In fact, many of these efforts are directed at incorporating the landscape approach into the mission of individual resource management agencies. Those efforts that are interagency collaborations largely focus on a smaller geographic region or on a limited number of target species.

Among this large number of programs, there are only a few other landscape-scale programs that combine science delivery with Landscape Conservation Design. Even within this sub-set of science delivery programs, LCCs have a distinct niche in that they identify and prioritize conservation science needs broadly; fund and otherwise support research projects that address these needs; and ensure that the results and products derived from these projects can apply to conservation efforts. They also operate regionally across the entire nation and across borders with Mexico and Canada, and they are not limited in scope to a few target species within a landscape-scale approach. Furthermore, the LCCs aim to develop a nationwide network of partnerships that play an important role working across jurisdictional boundaries and developing an integrated approach to conservation at the landscape scale. Although there is some overlap, the committee concludes that the extent of overlap is acceptable, given that there is usually a good rationale for the redundancy and good coordination with the overlapping partners. LCCs differ significantly in governance, geographic extent, mission, and activities from these other programs presented in this chapter, and hence are not fundamentally duplicative.

Of the programs reviewed, only the Joint Ventures and the CSCs appear to have some potential for redundancy that might need mitigation (see recommendations below). The committee found sufficient coordination between the NOAA RISAs and the LCCs, given different heritages, agency mandates, and governance structures. LCCs appear to benefit from the RISAs' long history, greater focus on social science and human communities, and capability of producing training events.

The CSCs and LCCs were initiated by Secretarial Order No. 3289 and intended to be complementary. According to their respective strategic plans, the CSCs aim primarily to develop science and the LCCs to develop strategic conservation priorities, Landscape Conservation Designs, and applied science to inform the conservation priorities. However, based on the committee's review of funded projects, both entities aspire to fund science, develop tools, and translate research for decision makers. Given that research priorities are developed for both the LCCs and CSCs by a committee of resource managers and other partners, it is not surprising that the resulting research might be similar and cause potential redundancies.

Recommendation: The LCC and CSC programs should be more clearly delineated. They should explicitly state how their research efforts differ and how they complement each other, identify and build on existing examples of coordination across the network, and make adjustments as appropriate. At the regional scale, LCC coordinators and CSC federal directors should coordinate their activities, including calls for proposals, as much as possible to avoid duplication of effort.

By design, the Joint Ventures and LCCs are both very similar FWS programs. Once the LCCs develop clear strategic priorities and focus on target species or some particular priorities, they will likely develop their own identity and will become more clearly distinguishable from the Joint Ventures. Despite the recognized need and unique niche for these multiple landscape conservation partnerships, the number of such efforts poses significant challenges to some partners whose active engagement is critical to achieving success (both for the LCCs and related programs). In particular, state agencies might not have sufficient staff resources to actively participate in all partnerships and will have to prioritize which partnerships are most likely to contribute to their own priorities.

Recommendation: DOI should review the landscape and habitat conservation efforts, especially the Joint Ventures and the LCCs, to identify opportunities for improved coordination between these efforts. Special consideration should be given to the limited capacity of state agency partners to participate in multiple efforts simultaneously.

Lastly, the committee was asked to consider "what may be gained or lost by consolidating [these FWS programs] in the LCC program." The committee concludes based on its analysis above that none of the programs would benefit from consolidation given their distinct roles in addressing the nation's conservation challenges.

6

An Assessment of the Early Accomplishments and Likely Long-Term Outcomes and Impacts of the Landscape Conservation Cooperatives Network

Congress asked the committee to assess "whether there have been measurable improvements in the health of fish, wildlife, and their habitats as a result of the program." This chapter addresses this congressional request and the U.S. Fish and Wildlife Service's (FWS's) charge to evaluate the following questions: What goals (and/or objectives) have been achieved? What improvements in managing and conserving habitats and fish and wildlife species might be reasonable to expect from the Landscape Conservation Cooperatives (LCCs) program in the timeframe it has existed? What longer-term impacts are likely to be realized?

The first section of the chapter provides examples of the goals and objectives that have been achieved so far. Given the youth of this program, the committee provides its explanation at the end of the chapter for why it is too soon to expect "measurable improvements in the health of fish, wildlife, and their habitats." Consequently, the committee reviewed examples of other landscape-scale conservation programs to glean an indication of the long-term outcomes that can be reasonably expected from the LCCs. The committee also examined analyses that identify design and implementation features of landscape-scale conservation efforts that appear to be correlated with eventual success and applied these to the LCCs.

EARLY ACCOMPLISHMENTS OF THE LCCs AND THE LCC NETWORK

As discussed in Chapter 1, the geographic scope of the LCC Network includes much of North America and the Pacific and Caribbean Islands. Therefore, to achieve national coverage, the LCC program was designed as a network consisting of 22 regional, self-directed cooperatives. To develop such a network of LCCs, the individual LCC partnerships had to be put in place first.

The creation of these 22 LCCs is an early achievement and an important process outcome. They were evaluated as part of the Science Investment and Accountability Schedule (SIAS) Activity Area titled "organizational operations" (see Chapter 4). For all 22 individual LCCs, the following steps have been achieved:

- Appointed staff coordinators and science coordinators
- Appointed steering committees
- Developed the governance structure
- Convened the steering committees to develop a common set of goals and articulate the common vision and goals
- Initiated or completed the development of a strategic plan and science priorities

As a result of achieving these process outcomes, the coordinators have begun to make progress toward the means objectives articulated by the SIAS: (1) engagement and coordination, (2) leveraging resources, and (3) engaging the technical community and technical staff.

The task of assessing "measurable improvements in the health of fish, wildlife, and their habitats" as well as "what goals have been achieved" is difficult to assess for several reasons: (1) the LCCs do not have the authority to manage fish, wildlife, and habitats; (2) therefore, measuring improvements is not possible due to the difficulty in attributing results to collaborative, conservation efforts such as the LCCs; (3) assessing which objectives have been achieved at the national program level is difficult because the LCCs' evaluation tool measures progress toward the Strategic Habitat Conservation Handbook (National Technical Assistance Team, 2008) goals and objectives instead of measuring progress toward the goals of the LCC Network Strategic Plan (hereafter referred to as the strategic plan; LCC, 2014); and (4) the LCC Network does not have an assessment tool or an effort to synthesize results from efforts across the network as a whole (see detailed description in Chapter 4).

Despite these challenges, the committee has attempted to describe some early progress by summarizing results from the individual LCC evaluations (i.e., results from the

SIAS assessments (see Chapter 4 for detailed descriptions). The committee provides some examples of other early accomplishments under each of the four strategic goal areas that correspond to the objectives listed in the strategic plan. Because the SIAS and the strategic plan are not aligned with regard to objectives and goals, the following should not be viewed as a formal, summative evaluation.

Goal 1: Conservation Strategy

> *Objective 1:* "Identify shared conservation objectives, challenges, and opportunities to inform landscape conservation at continental, LCC, island, and regional scales."

Almost all of the 22 LCCs have completed their strategic plans, and as such, have completed or initiated the identification and articulation of such shared conservation objectives. These plans appear to represent the shared objectives of the steering committee members of each LCC, assuming that each committee member was actively involved in the process. However, it is unclear how the broader stakeholder community was involved in the development of these shared objectives. The broad goals articulated by each LCC are consistent with and support the goals identified by the strategic plan. Based on these shared conservation objectives, the individual LCCs have funded projects jointly with their partners to guide how to address some of these conservation priorities. The SIAS results indicate that all of these strategic planning efforts have built on existing large-scale planning efforts, and some have effectively leveraged those existing efforts. For example, the Northwest Boreal LCC conducted a "comprehensive science and management information needs assessment." This needs assessment included broad stakeholder outreach to "identify priorities, information gaps, and current/future planning efforts." The LCC built its planning effort based on those results (SIAS NWB LCC FY 2014).

> *Objective 2:* "Develop then deliver (through partners) regional landscape conservation goals and designs."

Many important accomplishments can be listed for this objective. To date, the LCCs have funded 135 projects to conduct vulnerability assessments, and all LCCs have initiated the development of vulnerability assessments. Vulnerability assessments are a critical first step in adapting to the impacts of climate change (NRC, 2010; AFWA, 2012) and help resource managers identify priorities. Of the 22 LCCs, 4 report in their SIAS to have completed or adopted vulnerability or landscape assessments for 100 percent of the geography or 100 percent of the LCC's priority resources, and 13 report to have completed or adopted vulnerability or landscape assessments for at least 66 percent of the geography or at least 66 percent of the LCC's priority resources.

In addition, the LCCs have initiated 68 projects aimed to deliver conservation designs. Some LCCs have completed Landscape Conservation Designs (LCDs), such as the *Conservation Blue Print for the South Atlantic*, or the LCD in the Connecticut River Watershed (North Atlantic LCC). LCDs are defined in the strategic plan as: "[a]n iterative, collaborative, and holistic process that provides information, analytical tools, spatially explicit data, and best management practices to develop shared conservation strategies and to achieve jointly held conservation goals among partners." How these plans are implemented will determine whether ends objectives can be accomplished (see Appendix C, guidance on Landscape Conservation Design). LCDs are important outputs to guide and improve resource management.

Because of the importance of the LCC's LCD process to the LCC efforts, the committee reviews it in greater detail here. These LCDs are intended to "support resiliency and adaptation to both global change and regional landscape challenges, while ensuring the inclusion of all partners and stakeholders." The FWS Strategic Habitat Conservation Framework has served as the conceptual foundation for Landscape Conservation Design and planning (National Technical Assistance Team, 2008), and the adaptive management cycle from that framework has helped inform both LCDs and adaptive management more broadly in LCCs (see Figure 6.1). The Conservation Measures Partnership's *Open Standards for the Practice of Conservation* Version 3.0 (CMP, 2013) have also informed these designs (see South Atlantic LCC example below). The LCC Network Conservation Science Plan (LCC Science Coordinators Team, 2015) provides a step-by-step process for this planning and design, with the planning theme focused on establishing

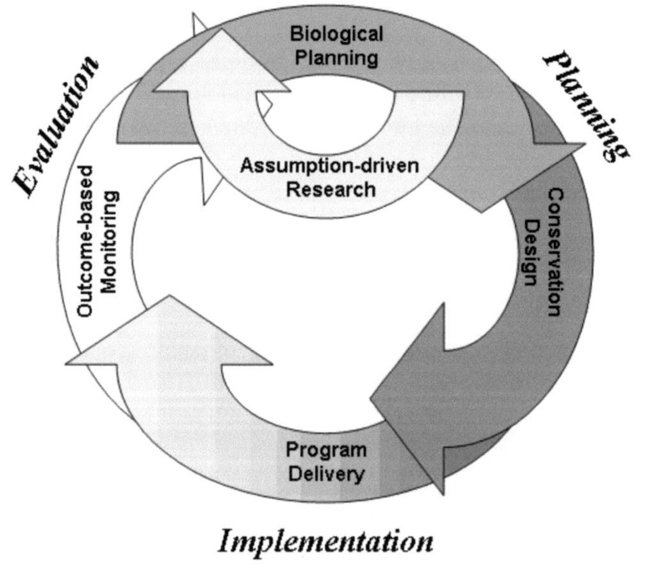

FIGURE 6.1 The basic Strategic Habitat Conservation cycle.
SOURCE: National Technical Assistance Team, 2008.

targets (ecological features) and goals for these targets in the face of scenarios of future change. The design theme turns those objectives and targets into a spatial network of ecologically connected conservation areas, including a threat assessment and the translation of goals into resource management objectives. The planning theme is intended to include ecological processes, ecosystem services, cultural resources, and climate adaptation planning. LCC science coordinators are attempting to bring consistency and compatibility to conservation targets in regionally adjacent LCC units. Ultimately, there is a network goal of creating "seamless and compatible" Landscape Conservation Designs within LCCs that "collectively contribute to an ecologically connected network of functional landscapes and seascapes," presumably across the United States.

The most advanced of these is the Conservation Blueprint of the South Atlantic LCC, where version 2.0 was released in June 2015 (see Figure 6.2). More than 400 individuals from more than 100 organizations were involved in preparing the blueprint. It was developed through a series of regional workshops where participants selected small watersheds as priority conservation areas and then assigned various conservation actions to those watersheds based on a standard set of actions derived from the *Open Standards for the Practice of Conservation* (CMP, 2013). Numerous existing conservation plans were used to develop the priorities for the Conservation Blueprint, including but not limited to The Nature Conservancy's Ecoregional Plans and several data sets that formed the basis for priority areas in State Wildlife Action Plans of the region. There is extensive documentation for how the map was prepared.[1] Intended uses of the blueprint are: "finding places to pool resources, raising new conservation dollars, guiding infrastructure development, developing conservation incentives, showing how local actions fit into a larger strategy, and locating places to build resilience to major disasters." Because of the coarseness of the data layers, the blueprint is not intended to be used for site-specific planning of conservation strategies and actions.

The North Atlantic LCC is also investing considerable effort in landscape conservation planning. Although it is not yet to the point of producing an LCD, its work grew out of the integrated planning efforts of the Northeast Fish and Wildlife Agencies' Regional Conservation Needs Program (Terwilliger Consulting, 2013). This program essentially synthesized the State Wildlife Action Plans of 13 northeastern states for the purposes of addressing regional conservation priorities, bringing consistency to the planning efforts across the region, highlighting what is most important in terms of wildlife conservation for the region, and organizing data and information for future efforts. Like the South Atlantic LCC, the North Atlantic LCC is developing a Conservation Planning Atlas, and all of their data and map layers are available on a mapping platform powered by Data Basin.

[1] See http://salcc.databasin.org/maps/a46404d870df478f871e1af23d8da539.

Conclusion: In reviewing the guidance in the LCC Network Science Plan and the Conservation Blueprint of the South Atlantic LCC, the committee finds some shortcomings. Although the basic steps described in the LCC Network Science Plan are useful and parallel the conservation science and planning literature to some degree, they fall short in several important ways (of what planners, scientists, and practitioners will need in LCCs to develop adequate landscape conservation plans that can be implemented in a manner to achieve conservation outcomes) (Groves and Game, 2015).

First, the existing guidance primarily focuses on the spatial aspects of planning and gives limited attention to the strategies and actions that will be needed to conserve places identified in these blueprints. By integrating spatial planning (where should conservation areas be located) with strategic planning (what strategies do we advance to conserve places), conservation planners and practitioners can better set priorities.

Second, the existing guidance mostly emphasizes the identification of places to focus attention on conservation targets (species, ecosystems) when it is likely that the conservation plans and LCDs of LCCs in the future will need to focus on multiple objectives. These objectives include conservation targets, ecosystem services, cultural resources, and other social and economic objectives that the LCC stakeholders may want to achieve in concert. Marine spatial planning is a great example of a method of planning for multiple objectives.

Third, the existing guidance needs more attention on understanding the socio-ecological systems in which any LCC conservation plan or LCD is being conducted, with a greater understanding of the social systems being critical to long-term success of any broad-scale conservation plan.

Finally, all conservation plans and LCDs face risks and uncertainties, and are more likely to succeed if these risks and uncertainties are formally acknowledged and accounted for in the planning process. Because of the wealth of guidance available for how to prepare conservation plans and Landscape Conservation Design, the committee has summarized some important lessons from recent conservation planning peer-reviewed literature that could help improve the overall methodology of landscape conservation planning and design being advanced by the LCCs (see Appendix C).

> *Objective 3:* "Integrate regional or other scale-specific conservation designs to align and focus conservation action at the network scale."

As described in detail in Appendix B, the Mississippi River Basin/Gulf Hypoxia Initiative is an example of aligning LCCs to focus conservation planning and potential actions at the network scale. Another example of working across LCCs is the Gulf Coast Vulnerability Assessment, where four LCCs are jointly evaluating the vulnerability of coastal habitat to sea level rise. In fact, the SIAS results

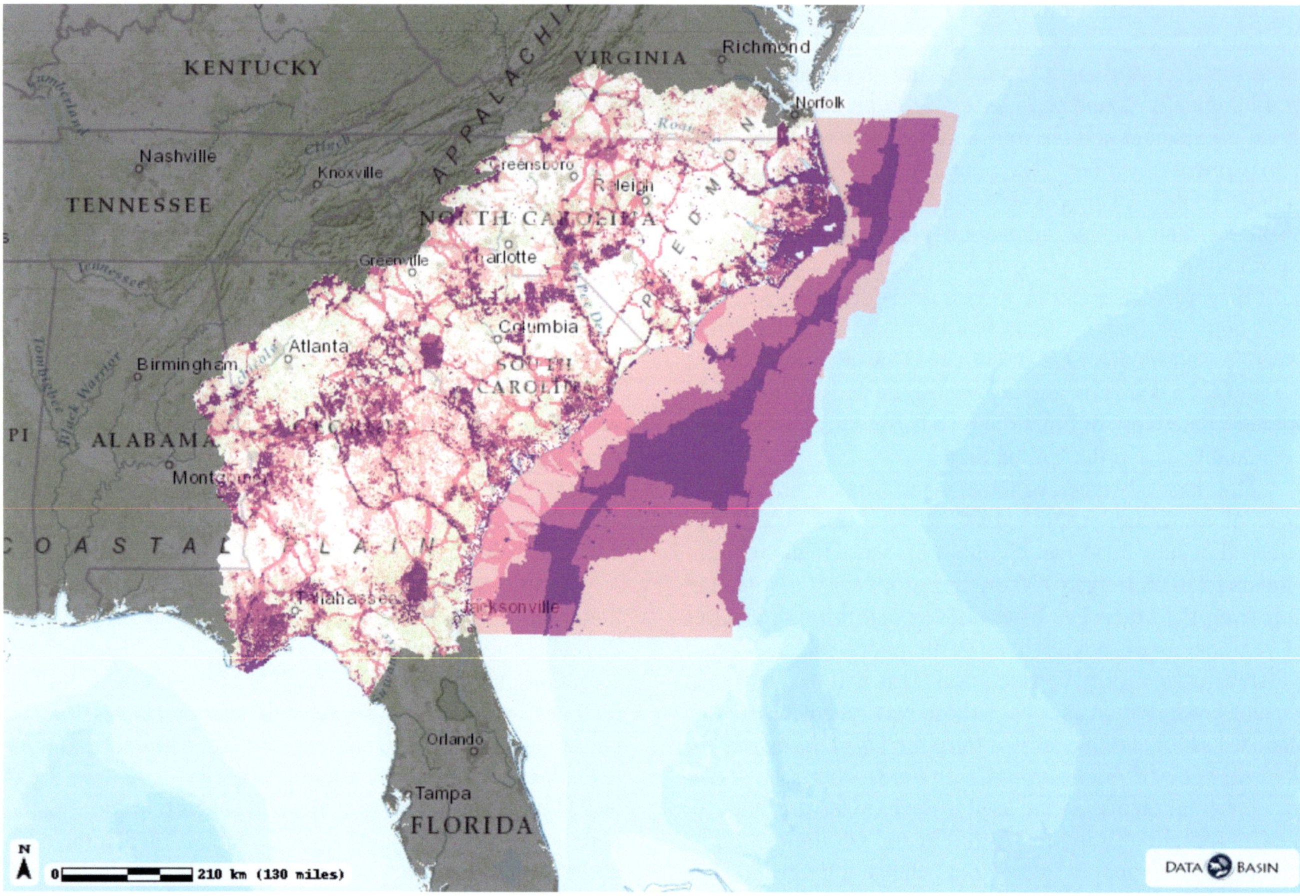

FIGURE 6.2 South Atlantic LCC Conservation Blueprint 2.0.
SOURCE: South Atlantic LCC, 2015.

indicate that all LCCs work with at least one other LCC and many report that "landscape-level conservation delivery has occurred as a direct result of" working across LCCs.

Goal 2: Collaborative Conservation

> *Objective 2:* "Identify and explore opportunities for collaborative actions within the LCC Network."

Several joint efforts that span multiple LCCs are in progress, such as the Mississippi River Basin/Gulf Hypoxia Initiative described in Appendix B and the effort to implement the Southeast Conservation Adaptation Strategy. All LCCs indicate in their SIAS that they collaborate with at least one other LCC, and the LCCs report to be "moderately or fully integrated" as part of the LCC Network. Thus, these results indicate progress has been made toward achieving this objective and work across boundaries. Although this goal highlights the LCC Network's emphasis on collaborative conservation, it is unclear what framework and structure are in place to facilitate work across LCC boundaries. On the SIAS benchmark that assesses whether the individual LCCs view themselves as functioning as "part of [the] integrated network of LCC partnerships," LCCs reported answers ranging from "moderately" to "fully."

> *Objective 3:* "Demonstrate, monitor, and evaluate the value and effectiveness of the LCC Network."

Some early progress toward monitoring and evaluating the value and effectiveness of the LCCs and the LCC Network has been noted: the FWS developed an assessment tool—the SIAS—to evaluate the individual LCCs. However, as discussed in Chapter 4, the ability of "demonstrating effectiveness" of the program is severely hampered by lack of an effective process to monitor and evaluate progress toward the LCC Network-wide goals as outlined in the strategic plan. For a detailed description, see Chapter 4.

Goal 3: Science

> *Objective 1:* "Identify shared science, information, and resource needs at the network scale."

Although individual LCCs have identified science needs and funded research, at the network scale, progress toward this objective appears nascent. A network-wide LCC Science Plan is being developed and is nearing completion.

> *Objective 2:* "Promote collaborative production of science and research—including human dimensions—as well as the use of experience and indigenous and traditional ecological knowledge among LCCs, Climate Science Centers (CSCs), and other interested parties; use these to inform resource management decisions, educate local communities, and address shared needs."

As discussed in greater detail in Chapter 5, collaboration and coordination between LCCs and CSCs is uneven across the regions and will continue to be a critical need. LCCs have funded and produced science for up to 3 years, and examples can be cited where such research and tool development has contributed to management of resources. As described in detail in Appendix A, a joint project between an LCC and AFWA resulted in more strategic fire management. A process needs to be developed that ensures that coordination at the individual LCC/CSC level translates to effective coordination at the network level as well.

> *Objective 3:* "Demonstrate and evaluate the value and improve the effectiveness of LCC science."

A large number of projects have been funded or completed to deliver science for the LCCs and their partners.[2] For example, 142 projects are listed under the "data acquisition and development" category, and 221 projects are listed under the "decision support" category. However, "demonstrating and evaluating" these individual projects at the LCC Network scale is challenging without an effective evaluation process. A few examples have been documented where LCC-funded research endeavors have improved conservation actions. For instance, the Appalachian LCC developed models and mapping tools to inform the "Assessing Future Energy Development Across the Appalachians" project. This plan was developed as a result of aforementioned research and intends to guide the balance between energy development and resource protection. Similarly, research funded by the Great Basin LCC is now enabling resource managers to more strategically allocate limited resources to battling wildfires that threaten critical sage-grouse habitat (Chambers et al., 2014).

Goal 4: Communications

> *Objective 1:* "Communicate the existence and application of LCC Network science, products, and tools to partners and stakeholders in a form that is understandable, publicly accessible, engaging, and relates to what matters to end users and society."

As identified during the most recent LCC Council meeting of March 26, 2015, communication is a critical objective and will be an ongoing effort. Examples have been documented of research and tools funded by individual LCCs to be communicated to stakeholders. For instance, the Aleutian and Bering Sea Islands LCC completed an analysis of major shipping routes in the Aleutian and the critical areas to avoid. This analysis and the areas to be avoided have been shared with the International Maritime Organization.

In summary, the above examples illustrate a range of early activities and accomplishments of the LCCs. This is not a comprehensive list and cannot do justice to the full range of accomplishments and milestones reached by the individual LCCs in the 4 years since the first LCCs were initiated. Yet, these examples indicate that diverse members of the LCC Network report progress in implementing key goals and objectives. It is beyond this study's scope to summarize the extent and wealth of accomplishments captured in the 22 SIASs submitted by the individual LCCs, however.

Conclusion

Given the short time since the LCCs and the LCC Network were established, these are the types of process milestones that are reasonable to expect from the LCCs, which have the potential to improve management of fish and wildlife in the future. A few examples of the tools developed have already led to improved resource management (see Appendix A, case study on sage-grouse), but it would be unreasonable to expect many such outcomes at this point because science and tool development projects alone typically take 2 to 3 years to complete before conservation actions can be set. As previously discussed, an evaluation process needs to be developed that can capture and synthesize the accomplishments of the individual LCCs to provide a network-wide assessment of the achievements.

WHAT LONGER-TERM IMPACTS HAVE RESULTED FROM OTHER LANDSCAPE-SCALE EFFORTS?

To better understand what long-term impacts might be realized by the work of the LCCs, the committee reviewed evaluations and outcomes of other landscape-scale conservation initiatives with a longer history than the LCCs. These are provided to demonstrate the viability of the landscape-scale approach and to identify some of the components (means and processes) that characterize successful landscape-scale

[2] See http://lccnetwork.org/projects.

efforts. Below, the committee discusses some accomplishments and key components of these apparently successful landscape approaches, but the committee cannot verify that there is a causal link between these components and the impacts. The committee anticipates that building a causal link between the collaborative, large-scale activities of the LCCs and concrete, positive impacts on biodiversity conservation will remain a difficult and elusive activity. The evaluations reviewed were for National Heritage Areas (NHAs), Pennsylvania Conservation Landscapes (PCLs), Yellowstone to Yukon (Y2Y), and the Atlantic Coast Joint Venture (ACJV). The first three examples were selected because we could draw from previous evaluations. The last effort was chosen because of the similarities to the LCCs (see Chapter 5 for more details on the similarity in program characteristics). Each of these initiatives was evaluated or reviewed using a different methodology and at different times in the program's life cycle. However, each initiative discusses how progress toward stated goals was achieved and provides some analysis of the collaborative management practices that were important to success in working on a landscape scale.

Each initiative and its assessment are summarized below.

National Heritage Areas

NHAs reflect the nation's significant and diverse heritage landscape. Forty-nine large landscape regions from Alabama to Alaska have been designated by Congress, ranging in size from smaller than a county to large bi-state regions. The first NHA was designated in 1984. The goals of NHAs are cultural and natural heritage conservation, interpretation, recreation, and community revitalization. Each area is locally managed, often by a coalition of partnership agencies and organizations with funding and technical support from the National Park Service (NPS). Each NHA has a management plan that sets forth the priorities for project implementation. Most receive federal operational funding and administer small grant programs.

The NPS undertook the evaluations of 12 of the long-established NHAs. These reviews could be characterized as summative evaluations examining the effectiveness and outcomes of the selected areas after 20 or more years of operation. In general, the evaluations reported positive findings. All but one of the NHAs addressed and made progress on each of the conservation goals identified in the area's legislation and approved management plans. Based on a review of funding allocation directed to completed projects, the 12 NHAs focused on the goals of cultural and natural resource conservation (31 percent) and education and interpretation (26 percent); other work included recreational development and heritage tourism activities. The evaluations of this work stated that the individual NHAs had fulfilled or successfully fulfilled the area's resource conservation goals. The evaluations also concluded that the work was carried out with a high level of partnership and citizen engagement. The evaluations documented the development of network management strategies and the ability to leverage funding for project development. Finally, the reviews noted the importance of continued NPS funding and support to the sustainability of the work (Barrett, 2013).

Pennsylvania Conservation Landscapes

The Pennsylvania Department of Conservation and Natural Resources (DCNR) has focused its efforts on seven regions in the state with significant public lands. The Conservation Landscape program's goals are sustainability, conservation, community revitalization, and recreational development. DCNR provides substantial funding for operations and a regrant program. However, each of the seven initiatives is managed by a local steering committee with a dedicated staff person provided by one of the local partners. Local partners set the priorities for the landscape. The program was initiated in 2004.

DCNR undertook the evaluations of PCLs early in the development process after 5 years (formative evaluation) to improve and inform the direction of the program and to provide a possible justification for the program with a coming change in administration. The evaluation report concluded that partnership development of the two most mature PCLs showed significant progress in long-term stewardship of public lands and the development of a positive relationship with adjacent communities. The leadership and financial commitment of DCNR was found to be critical to the future success of the program (Patrizi et al., 2009). (Note: The program was continued by the next administration.)

Yellowstone to Yukon

Y2Y is a joint Canadian-U.S. initiative with the goal of preserving and maintaining the wildlife, native plants, wilderness, and natural process of the mountain ecosystem stretching more than 2,000 miles from Yellowstone National Park to the Yukon Territory. The Y2Y region includes two countries, five U.S. states, two Canadian provinces, two Canadian territories, the reservation or traditional lands of more than 30 Native governments, and a number of government land agencies. Y2Y offers science-based education and stewardship programs that encourage conservation of the area's natural resources. Funding comes from a mix of public and private sources, and it was a challenge to find support in the early days of the project. This initiative was launched in 1994.

As part of the 20-year anniversary, a report titled *The Yellowstone to Yukon Vision* was prepared by the nonprofit organization that manages the overall project. The report summarizes the multiple habitat improvement projects between 1993 and 2013 as part of this large, transboundary effort. The report lists the strength of the network that implemented the Y2Y vision as an important component that contributed to its success. However, it also noted the challenge of a nongovernmental organization in sustaining the project until it could show results.

Atlantic Coast Joint Venture

The ACJV is one of many Joint Ventures (see also Chapter 5); and it is focused on the conservation of habitat for the native birds of the Atlantic Flyway of the United States from Maine to South Florida. The ACJV was originally formed—as were others—as a regional partnership under the North American Waterfowl Management Plan (NAWMP) in 1988. This particular Joint Venture includes all the states along the Atlantic Coast and Puerto Rico, and partners from federal agencies (FWS, U.S. Department of Agriculture, Forest Service, U.S. Geological Survey, and NPS), and nongovernmental organizations including American Bird Conservancy, Ducks Unlimited, The National Audubon Society, The Nature Conservancy, and the National Fish and Wildlife Foundation. The ACJV's structure (see Figure 6.3) connects science with management through the formation of technical committees that inform particular initiatives. The steering committee coordinates the work of the various technical committees and the managing board develops, approves, and implements the ACJV's implementation plan. The implementation occurs at a smaller scale either through state working groups or through focus area working groups. Partners in the ACJV are expected to contribute funds and activities that advance their jointly developed goals. Progress was measured primarily through the number of acres of habitat conserved.

The ACJV was not evaluated by external review, but the ACJV itself updated its strategic and implementation plans to include an assessment of progress made. Furthermore, the Joint Venture program as a whole was described by Giocomo et al. (2012). As of 2005, the ACJV had managed to conserve (through protection, restoration, and enhancement) almost 3 million acres of wetland since the inception of this Joint Venture in 1988. Future assessments and objectives will attempt to develop habitat conservation goals that are biologically linked to the breeding population goals.

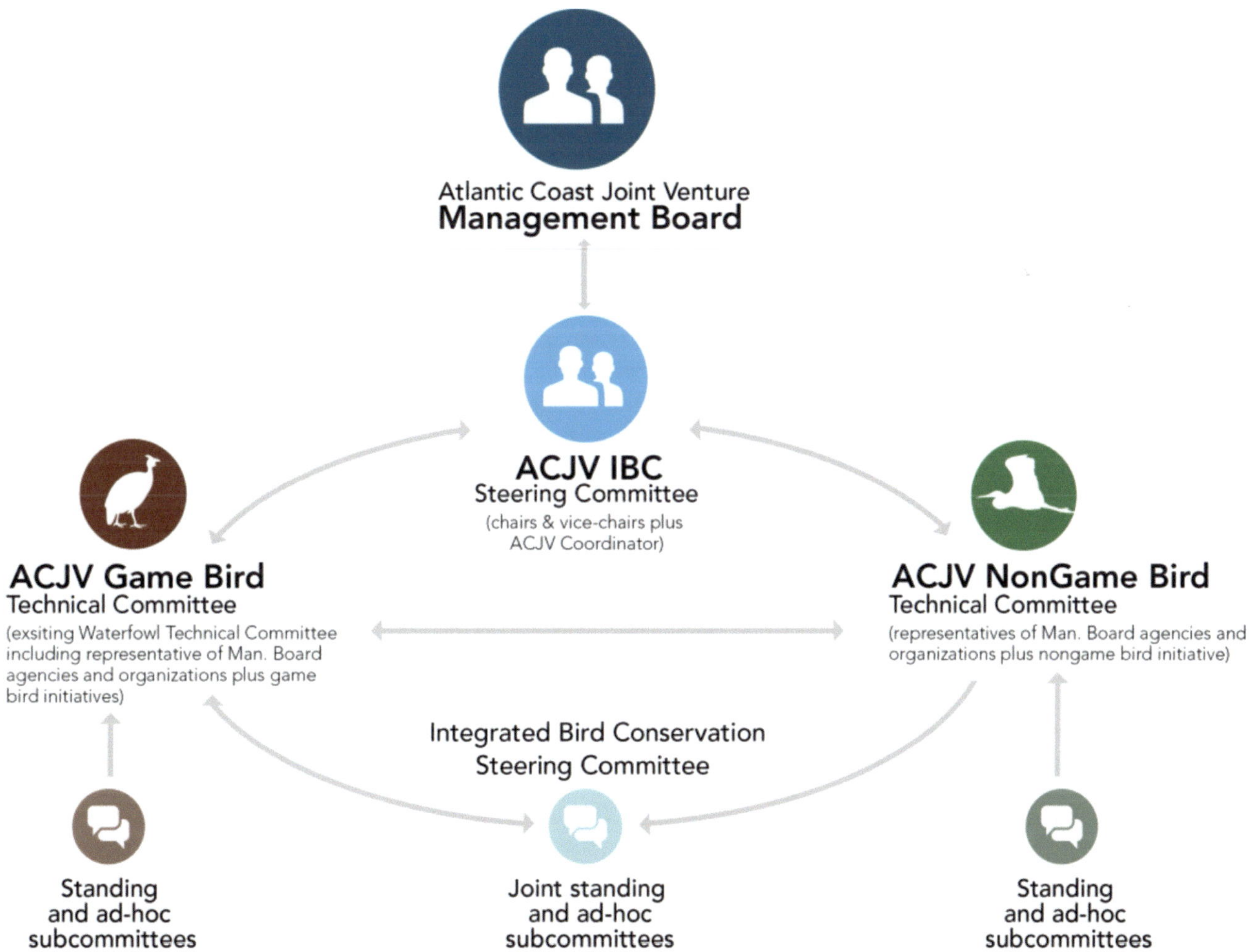

FIGURE 6.3 Schematic of the governance structure for the Atlantic Coast Joint Venture.
SOURCE: http://acjv.org/about-us/acjv-structure.

In summary, there are significant differences in the goals and outcomes between the landscape-scale efforts of the NHAs and PCLs and the LCCs. NHAs, PCLs, and Y2Y take place at a smaller scale and in some cases are more focused on cultural conservation and community development goals. The Joint Venture program's geographic focus is identical to that of the LCCs, having both continental-scale goals and a regional focus. However, in contrast to the LCCs, the ACJV works on more targeted species (i.e., migratory birds). Despite differences in intent, scale, and focus, there are similarities that may help inform the potential for long-term impacts of landscape-scale programs such as the LCCs. Specifically, LCCs are charged with the process of forming a network of partners to address issues on a landscape scale. For this reason, it may be useful to identify some of the components of landscape-scale projects and programs that can contribute to achieving programmatic goals.

COMPONENTS OF A LANDSCAPE-SCALE INITIATIVE IMPORTANT TO YIELDING DESIRED LONG-TERM OUTCOMES

It is challenging for the committee to address the statement of task question "What long-term impacts are likely to be realized?" because it is difficult to predict the future support for the LCC Network and individual LCCs, as well as many other factors that will determine whether the LCCs' goals can be achieved. Thus, in this section the committee discusses a number of critical elements and processes that have been identified as important to the effectiveness of both the above examples and to the emerging field of large-landscape conservation.

Examining these components may be useful in projecting potential outcomes and successes of the LCCs. As discussed in Chapter 2, there is a substantial literature on landscape approach, conservation design, and collaborative governance; however, the literature evaluating collaborative governance and landscape-scale conservation practice is more limited. Recent publications have identified a continuum of approaches to working on regional collaborations and elements important to landscape-scale conservation (McKinney and Johnson, 2009; McKinney et al., 2010; Curtin 2015). Laven and others (Martin-Williams, 2007; Laven et al., 2010) have used data from NHA evaluations to explore the factors that have sustained the effectiveness of that program over time in a large-landscape setting. Curtin (2015) examines the underpinning of large-landscape work, including theories of collective impact, distributed cognition, and innovation and adaptation. Drawing from this work and the four case studies, the committee highlights some of the key components and how they are addressed as part of the LCCs.

A Unifying Theme or Story

A common vision solidifies the commitment of partnerships that have different perspectives. The vision can be based on either shared future desired state or shared problems. A common understanding provides stability in an inherently fragile system (Martin-Williams, 2007). This was identified as an important factor in operationalizing the NHAs serving as the "glue" that holds the areas together (Laven et al., 2010). In the Y2Y anniversary report, the vision for species conservation on a bi-national scale was identified as an important factor. It was the unprecedented big idea that garnered a great deal of support and momentum (Yellowstone to Yukon Conservation Initiative, 2014). In the case of the Joint Ventures, the NAWMP and its central focus on waterfowl habitat protection provides the "glue" that allows the self-directed regional Joint Ventures to contribute in their own way to the common goal of conserving habitat, and it provides the central focus and boundaries around the scope of the work.

The stated purpose of the LCCs is responding to climate change and other landscape-scale stressors on land, water, and other natural and cultural resources, and this rationale could serve as the driver for the work of the individual LCCs.

Conclusion: This vision may be too broad and not as compelling as a place-based or species-based initiative, for example, the Y2Y's objective to protect an indicator species such as the grizzly bear. Similarly, the Joint Ventures' focus on waterfowl has provided them a central theme that helps justify the geographic boundaries and helps set priorities at a sub-regional level. However, the vision and goals of the individual LCCs might be more specific and defined by the partners. As a result, the priorities identified by the individual LCC steering committees might provide sufficient "glue" to compel collaborations.

Partnership and Network Development

Strong stakeholder engagement and an expanding network of partners is a critical element in large-landscape practice (McKinney et al., 2010). Evaluations determined that NHAs activated a network of partners from the national, state, and local sectors and that this network was a significant factor in the effectiveness of these regional efforts (Laven et al., 2010). In fact, some of the NHAs had more than 100 partners. NHA partnerships also promoted the development of an intergovernmental domain (Martin-Williams, 2007). Listening to the public and building trust and partnership relationships in the PCLs helped preserve resources and build social capital for the initiative (Patrizi et al., 2009). While there is no formal evaluation of the Joint Ventures' network strength, Giocomo et al. (2012) refer to the Joint Ventures as a highly successful partnership for bird conservation at a continental scale. During the information-gathering process for this study, comments from a number of stakeholders supported the view that Joint Ventures are a successful partnership. In the case of the ACJV, many partners of the management board have contributed to habitat conservation in significant ways. However, without a formal evaluation, it is not possible to determine whether these conservation actions were a direct result from the collaborative decision making as part of the management board or the ACJV steering committee.

LCCs were created explicitly using a partnership model to better integrate science and management, and to ensure an effective network in 22 regional landscapes. The steering committees for LCCs have members from federal, state, tribal, nongovernmental organizations, and other organizations, associations, or industries with an interest in landscape-scale conservation.

Conclusion: Although the design of the program indicates the intent to develop strong stakeholder engagement, as discussed in Chapter 4, it will require an assessment of the range of partners and the quality and functionality of these networks to indicate the potential for future success.

Adaptive Management

Adaptive management is important because a collaborative conservation approach needs to adapt to meet changing stakeholder needs (Barrett, 2013; see Chapter 2), as well as changing environmental threats. Learning from partnership development among the PCLs was an important strategy for the PCL program's future (Patrizi et al., 2009). Giocomo et al. (2012) point to adaptive management as a central component of the Joint Ventures, which enables a seamless inclusion of science in resource management.

Conclusion: Monitoring the effectiveness of the LCC Network is a stated objective of the program. The strategies of the different LCCs to achieve their ends could be reviewed for how well they adjust to changing issues and respond to the needs of project partners.

Planning Documents

A key element in regional collaboration is a plan to move a project from vision to action (McKinney et al., 2010). For NHAs, their statutorily required management plans served as a roadmap for working at a landscape scale, and the goals in the NHAs' plans were used as benchmarks to measure progress (Barrett, 2013). Joint Ventures addressed this important element by establishing a management board that oversees the development of an implementation plan (ACJV, 2005). The ACJV also supports the development of partnerships at the scale of the conservation deliverable that was identified through the ACJV, whether at the state level or at a smaller, more focused scale. There is a process in place that reduces broad continental-scale goals to a scale appropriate for conservation delivery, based on a solid biological foundation (ACJV, 2009).

One of the purposes of the LCCs is to develop shared landscape-level conservation objectives. Each of the LCCs has established an initial mission and strategic plan for its region. At this point, it is unclear how they will move from these high-level strategic plans toward implementation of conservation activities, as well as what authority or financial capacity LCCs have to advance their vision. In the future, these strategic plans and associated implementation plans will be important for evaluating and measuring progress, as discussed in greater details in Chapter 4. Although LCCs do not have direct management authority, many of the partners on the steering committee do have such authority.

Conclusion: Moving beyond planning toward conservation delivery will depend on the LCCs' catalyzing such conservation actions by their steering committee members with such management authority.

Aggregating Project Impact

The aggregation of data into similar categories from the 12 NHA evaluations, using expenditures, financial leverage, and project completion, demonstrated the impact of the initiative (Barrett, 2013). For example, the Y2Y has produced a report that sums the increase in areas with protected designation within the project's landscape. The PCLs used total acres acquired (in total 66,000 acres were acquired between 2003 and 2008) to measure and aggregate the program's impact. The ACJV's updates on its strategic plan (ACJV, 2009) and its implementation plan (ACJV, 2005) include an assessment on how the partners have contributed to conservation of wetlands.

As discussed in the previous chapter, the LCCs could use the information collected for the individual 22 LCCs in aggregation to track progress in categories such as climate adaptation or objectives that address projects such as sage-grouse habitat or Mississippi River/Gulf of Mexico hypoxia (see Appendix B). Further details on the evaluation process are discussed in Chapter 4. However, as discussed in Chapter 4, the existing evaluation process does not enable easy aggregation of project impacts.

Conclusion: As discussed in Chapter 4, the ability of the LCC Network to demonstrate its overall impact would be greatly enhanced with better synthesis and aggregation of individual LCCs progress.

Significance of Leverage

This information can be used to show progress toward project goals in a more cost-effective manner (Barrett, 2013). NHAs were successful in leveraging funding and exceeded their 50 percent matching goal. The program managed to leverage funding for heritage infrastructure up to a 4:1 ratio (Barrett, 2013). The Y2Y anniversary report shows growth of the network and the power of leveraging support to achieving conservation outcomes such as habitat protection. The Joint Ventures' main program website states that "[o]ver the course of [the program's] history, Joint Venture partnerships have leveraged every dollar of Congressional funds 34:1 to help conserve 22 million acres of essential habitat for birds and other wildlife."

The LCCs' SIAS is tracking the leveraging support from regional parties as a proxy measure to demonstrate the strength of the partnership, as well as to show the program's impacts or ends achieved by working in partnerships (see additional details in Chapter 4). So far, most LCCs report the leveraging at greater than 67 percent. Because the LCCs themselves have only limited implementation authority and rely mainly on the participating partner organizations for implementing their strategic plans, tracking on-the-ground conservation actions as part of leveraging will be important to demonstrate progress. Given that they use a partnership model very similar to the Joint Ventures, it is likely that the LCCs would leverage their resources at a relatively high rate.

Role of Governmental Agencies

NPS—by convening and funding NHAs—played a critical role in providing the environment that fostered a broad coalition across many organizations. In this way, the agency furthered its goals to create a partnership to conserve nationally significant resources (Martin-Williams, 2007). The hands-on leadership of Pennsylvania's DCNR was an essential factor in the success of the Pennsylvania Conservation Landscape (Patrizi et al., 2009). There is evidence that if NPS withdraws support from NHAs, the program will be severely diminished or will not survive (Alliance of National Heritage Areas, 2013). As discussed in Chapter 5, the North American Waterfowl Management Plan preceded the Joint Ventures, which were established to implement the North American Waterfowl Management Plan. Congress supported the North American Waterfowl Management Plan by passing the North American Wetlands Conservation Act in 1988. This statute was reauthorized in 2002 and 2006 with expansions to include all habitats and birds associated with wetlands and funds up to $75 million per year. This support plays an important role in bringing partners together and providing the Joint Ventures with the financial resources to accomplish their goals.

The U.S. Department of the Interior authorized the creation of the LCCs with Secretarial Order No. 3289 (see Chapter 1 and Appendix E), which provides operating funds and grants and serves as a platform for interagency cooperation. Funding is administered through the FWS regional offices. As a result, LCCs are perceived by some as competing for funds available to other FWS programs. The agency and the FWS regional offices' continued support, including funding, will be an important signal to the partners about long-term viability.

CONCLUSIONS

Early Accomplishments

At this point, it is too early to expect "measurable improvements in the health of fish, wildlife, and their habitats" for three reasons. First, the LCCs do not have the authority to deliver conservation, but instead work to accomplish the goal of improved management through partners. It requires time to develop partnerships, to establish shared goals, and eventually for partners with the authority for conservation delivery to implement those goals. Only at that point can one expect to begin to see improvements in the health of target species or habitats.

Second, in addition to the time lag between program inception and conservation improvements on the ground, it will be difficult to measure these improvements. In the case of collaborative conservation, it will be difficult to apportion credit for how the LCCs or the many individual partners have contributed to a given outcome (see Appendix A, case study on sage-grouse, and discussion in Chapter 4 about metrics). Our examples of similar programs demonstrate that this linkage can be established, but the value-added of the convener is difficult to quantify or attribute.

Third, the LCC Network has not yet developed a process to aggregate accomplishments of the individual LCCs to a network-wide programmatic assessment.

Nevertheless, the committee found many early accomplishments for almost all of the 19 objectives, such as the identification of shared conservation objectives, vulnerability assessments on a large number of resources across the United States, development of Landscape Conservation Designs, and production and delivery of many research projects, as discussed in detail above. In fact, some tools and research results funded by individual LCCs have already been noted as improving resource management decisions (see Appendix A). These early accomplishments are in line with the types of process objectives the committee would expect to see achieved during the early inception phase of a new federal program of this scale.

Likely Long-Term Outcomes

The LCC Network is a young program with an ambitious set of goals. Delivering on such broad and ambitious goals will require that the individual, self-directed LCCs succeed at setting strategic priorities and identifying conservation deliverables, as well as at establishing a process that can result in the desired improvements in managing natural and cultural resources. As discussed in Chapters 1 and 2, the threats and challenges to the nation's natural and cultural resources require a landscape-scale approach to conservation, and it appears from the committee's review in Chapter 5 that no other federal program is in a position to meet this need.

As outlined above, important lessons can be learned from other landscape conservation programs that have existed much longer. Critical components that are important for such collaborative efforts include a unifying theme, strong stakeholder engagement, adaptive management, strategic planning efforts, metrics to aggregate project impacts, leveraging, and a lead agency that provides resources and leadership. Based on the discussion above, the LCCs have most of these components in place. As discussed in Chapter 3, the overall LCC Network's goals, structure, and functions are consistent with the landscape approach, including the components outlined above, and therefore it should be in a position to deliver on its long-term goal of improving cultural and natural resource management. However, a firm financial commitment seems essential to sustaining the LCCs.

If sustained and successful, the LCCs will provide a process by which stakeholders can engage at the landscape scale to set strategic conservation priorities that can span interest groups and narrow disciplinary or sectoral approaches. It will also provide an important body of knowledge and tools to improve resource management. However, the committee concludes that it would require the LCCs to develop a process that can account and track how their planning efforts result in the implementation of on-the-ground conservation. In previous chapters, the committee has provided specific recommendations that would further improve the ability of the LCC Network to deliver its vision of "[l]andscapes capable of sustaining natural and cultural resources for current and future generations." In summary, the committee concludes that the LCC Network has the required elements to contribute and add value to the nation's conservation challenge at the landscape scale.

References

Abell, R.A., D.M. Olson, E. Dinerstein, P.T. Hurley, J.T. Diggs, W. Eichbaum, S. Walters, W. Wettengel, T. Allnutt, C.J. Loucks, and P. Hedao. 2000. Freshwater Ecoregions of North America: A Conservation Assessment. Washington, DC: Island Press.

ACCCNRS (Advisory Committee on Climate Change and Natural Resource Science). 2015. Report to the Secretary of the Interior. Available at https://nccwsc.usgs.gov/sites/default/files/files/ACCCNRS_Report_2015.pdf (accessed October 5, 2016).

ACJV (Atlantic Coast Joint Venture). 2005. Atlantic Coast Joint Venture Waterfowl Implementation Plan. Available at http://acjv.org/planning/waterfowl-implementation-plan (accessed October 5, 2016).

ACJV. 2009. Atlantic Coast Joint Venture Strategic Plan, Updated. Available at http://acjv.org/documents/ACJV_StrategicPlan_2009update_final.pdf (accessed October 5, 2016).

Agee, J.K., and D.J. Johnson, Eds. 1989. Ecosystem Management for Parks and Wilderness. Seattle: University of Washington Press.

Allen, C.R., and L.H. Gunderson. 2011. Pathology and failure in the design and implementation of adaptive management. Journal of Environmental Management 92(5):1379-1384.

Alliance of National Heritage Areas. 2013. An Evaluation of 12 National Heritage Areas. Available at http://www.nationalheritageareas.us/pdf/Final-NHA.pdf (accessed October 5, 2016).

Axelrod, R. 1984. The Evolution of Cooperation. Princeton, NJ: Princeton University Press.

Axelrod, R. 1997. The Complexity of Cooperation: Agent-Based Models of Competition and Collaboration. Princeton, NJ: Princeton University Press.

Bagstad, K.J., D.J. Semmens, S. Waage, and R. Winthrop. 2013. A comparative assessment of decision-support tools for ecosystem services quantification and valuation. Ecosystem Services 5:27-39.

Ban, N.C., M. Mills, J. Tam, C.C. Hicks, S. Klain, N. Stoeckl, M.C. Bottrill, J. Levine, R.L. Pressey, T. Satterfield, and K.M.A. Chan. 2013. A social-ecological approach for conservation planning: Embedding social considerations. Frontiers in Ecology and the Environment 11:194-202.

Barrett, B. 2013. National Heritage Areas: Evaluating past practice as a foundation for the future. Presented at Fabos Conference on Greenways, University of Massachusetts at Amherst. Available at http://livinglandscapeobserver.net/wp-content/uploads/2013/04/Fabos-paper-2013-Barrett.pdf (accessed September 8, 2016).

Baylis, K., J. Honey-Rosés, J. Börner, E. Corbera, D. Ezzine-de-Blas, P.J. Ferraro, R. Lapeyre, U.M. Persson, A. Pfaff, and S. Wunder. 2015. Mainstreaming impact evaluation in nature conservation. Conservation Letters 9(1):58-64. doi: 10.1111/conl.12180.

Berger, J., and S. Cain. 2014. Moving beyond science to protect a mammalian corridor. Conservation Biology 28:1142-1150.

Berkes, F., and C. Folke, Eds. 1998. Linking Social and Ecological Systems: Management Practices and Social Mechanisms for Building Resilience. New York: Cambridge University Press.

Betts, M.G., L. Fahrig, A.S. Hadley, K.E. Halstead, J. Bowman, W.D. Robinson, J.A. Wiens, and D.B. Lindenmayer. 2014. A species-centered approach for uncovering generalities in organism responses to habitat loss and fragmentation. Ecography 37(6):517-527.

Binder, L.W., H. Campbell, G. Carbone, M. Finucane, G. Garfin, H. Hartmann, K. Ingram, D. Kenney, N. Lewis, C. Simpson, S. Trainor, B. Udall, and R. Webb. 2009. RISA 2020: A RISA Community Vision for the Future Regional Integrated Sciences and Assessments (RISA) Efforts to Match Advances in Climate Impacts Science with the Needs of Resource Managers and Planners. Available at http://cpo.noaa.gov/sites/cpo/RISA/RISA%202009%20Vision%202020.pdf (accessed October 5, 2016).

Bi-State Technical Advisory Committee Nevada and California. 2013. Bi-State Action Plan—Past, Present, and Future Actions for Conservation of the Greater Sage-Grouse Bi-State Distinct Population Segment. Prepared for the Bi-State Executive Oversight Committee for Conservation of Greater Sage-Grouse, March 15, 2012.

Bodin, Ö., B. Crona, M. Thyresson, A.-L. Golz, and M. Tengö. 2014. Conservation success as a function of good alignment of social and ecological structures and processes. Conservation Biology 28:1371-1379.

Burgman, M.A. 2005. Risks and Decisions for Conservation and Environmental Management. Cambridge, UK: Cambridge University Press, p. 488.

Buzbee, W.W. 2003. Recognizing the regulatory commons: A theory of regulatory gaps. Iowa Law Review 89:1-64.

Cash, D.W., W.C. Clark, A. Alcock, N.M. Dickson, N. Eckley, D.H. Guston, J. Jager, and R.B. Mitchell. 2003. Knowledge systems for sustainable development. Proceedings of the National Academy of Sciences of the United States of America 14:8086-8091.

Chambers, J.C., D.A. Pyke, J.D. Maestas, M. Pellant, C.S. Boyd, S.B. Campbell, S. Espinosa, D.W. Havlina, K.E. Mayer, and A. Wuenschel. 2014. Using Resistance and Resilience Concepts to Reduce Impacts of Invasive Annual Grasses and Altered Fire Regimes on the Sagebrush Ecosystem and Greater Sage-Grouse: A Strategic Multi-Scale Approach. General Technical Report RMRS-GTR-326. Fort Collins, CO: U.S. Department of Agriculture, Forest Service, Rocky Mountain Research Station. 73 pp.

Clark, T.W., E.D. Amato, D.G. Whittemore, and A.H. Harvey. 1991. Policy and programs for ecosystem management in the Greater Yellowstone ecosystem: An analysis. Conservation Biology 5:412-422.

CMP (Conservation Measures Partnership). 2013. Open Standards for the Practice of Conservation. Version 3.0. Available at http://cmp-openstandards.org/wp-content/uploads/2014/03/CMP-OS-V3-0-Final.pdf (accessed October 5, 2016).

Cook, C.N., M. Hockings, and R.W. Carter. 2010. Conservation in the dark? The information used to support management decisions. Frontiers in Ecology and the Environment 8(4):181-186.

Cook, C.N., M.B. Mascia, M.W. Schwartz, H. P. Possingham, and R.A. Fuller. 2013. Achieving conservation science that bridges the knowledge-action boundary. Conservation Biology 27(4):669-678.

Council on Climate Preparedness and Resilience, Climate and Natural Resources Working Group. 2014. President's Priority Agenda: Enhancing the Climate Resilience of America's Natural Resources, p. 16, footnote 14.

Crowder, L.B., G. Osherenko, O.R. Young, S. Airamé, E.A. Norse, N. Baron, J.C. Day, F. Douvere, C.N. Ehler, B.S. Halpern, S.J. Langdon, K.L. McLeod, J.C. Ogden, R.E. Peach, A.A. Rosenberg, and J.A. Wilson. 2006. Resolving mismatches in U.S. ocean governance. Science 313:617-618.

Curtin, C. 2015. The Science of Open Spaces: Theory and Practice for Conserving Large, Complex Systems. Washington, DC: Island Press.

De Leo, G.A., and S. Levin. 1997. The multifaceted aspects of ecosystem integrity. Conservation Ecology 1(1):3. Available at http://www.consecol.org/vol1/iss1/art3 (accessed October 5, 2016).

Doherty, K., J.D. Tack, J.S. Evans, D.E. Naugle, 2010. Mapping Breeding Densities of Greater Sage-Grouse: A Tool for Range-wide Conservation Planning. BLM Completion Report: Interagency Agreement # L10PG00911. Bureau of Land Management.

DOI (U.S. Department of the Interior). 2011. Landscape Conservation Cooperatives and Climate Science Centers Implementation Guidance. Available at: https://lccnetwork.org/resource/lcc-and-climate-science-center-implementation-guidance (accessed September 8, 2016).

Doremus, H. 2011. Adaptive management as an information problem. North Carolina Law Review 89:1455-1498.

Dunning, J.B., B.J. Danielson, and H.R. Pulliam. 1992. Ecological processes that affect populations in complex landscapes. Oikos 65(1):169-175.

Eigenbrod, F., S.J. Hecnar, and L. Farhig. 2011. Sub-optimal study design has major impact on landscape-scale inference. Biological Conservation 144:298-305.

Ellis, E.C., K.K. Goldewijk, S. Siebert, D. Lightman, and N. Ramankutty. 2010. Anthropogenic transformation of the biomes, 1700 to 2000. Global Ecology and Biogeography 19:589-606.

Fischer, J., D.B. Lindenmayer, S. Barry, and E. Flowers. 2005. Lizard distribution patterns in the Tumut Fragmentation "Natural Experiment" in south-eastern Australia. Biological Conservation 123:301-315.

Foley, J.A., R. DeFries, G.P. Asner, C. Barford, G. Bonan, S.R. Carpenter, F.S. Chapin, M.T. Coe, G.C. Daily, H.K. Gibbs, J.H. Helkowski, T. Holloway, E.A. Howard, C.J. Kucharik, C. Monfreda, J.A. Patz, I.C. Prentice, N. Ramankutty, and P.K. Snyder. 2005. Global consequences of land use. Science 309:570-574.

Forman, R. 2000. Estimate of the area affected ecologically by the road system in the United States. Conservation Biology 14(1):31-35.

Frampton, G. 1996. Ecosystem management in the Clinton Administration. Duke Environmental Law & Policy Forum 7:39-48.

Franklin, J.F. 1993. Preserving biodiversity: Species, ecosystems, or landscapes? Ecological Applications 3(2):202-205.

Freeman, J., and J. Rossi. 2012. Agency coordination in shared regulatory space. Harvard Law Review 125:1131-1211.

Freitag, M., and I. Winkler. 2001. Development of cooperation in regional networks: Mechanisms of coordination and support measures. In Collaborative Strategies and Multi-Organizational Partnerships, edited by Tharsi Taillieu. Leuven, Belgium: Grant, pp. 67-72.

FWS (U.S. Fish and Wildlife Service). 2010. Endangered and threatened wildlife and plants; 12 month findings for petitions to list the greater sage-grouse (*Centrocercus urophasianus*) as threatened or endangered. Federal Register 75:13909-14014.

FWS. 2013a. A Landscape-Scale Approach to Refuge System Planning. USFWS Final Report, Washington, DC: FWS. Available at http://www.fws.gov/refuges/vision/pdfs/PlanningGuideRev10.pdf (accessed September 8, 2016).

FWS. 2013b. Greater Sage-Grouse (*Centrocercus urophasianus*) Conservation Objectives: Final Report, February. Denver, CO: FWS.

FWS and Environment Canada, Canadian Wildlife Service. 1986. North American Waterfowl Management Plan: A Strategy for Cooperation (May 1986).

Gallant, A.L., T.R.Whittier, D.P. Larsen, J.M. Omernik, and R.M. Hughes. 1989. Regionalization as a Tool for Managing Environmental Resources. EPA/600/3-89/060. Corvallis, OR: U.S. Environmental Protection Agency, Environmental Research Laboratory. 152 pp.

Game, E.T., P. Kareiva, and H.P. Possingham. 2013. Six common mistakes in conservation priority setting. Conservation Biology 27(3):480-485.

GAO (U.S. Government Accountability Office). 2011. Performance Measurement and Evaluation: Definitions and Relationships. GAO-11-646sp. Available at gao.gov/new.items/d11646sp.pdf (accessed October 5, 2016).

Garton, E.O., J.W. Connelly, J.S. Horne, C.A. Hagen, A. Moser, and M.A. Schroeder. 2011. Greater sage-grouse population dynamics and probability of persistence. Studies in Avian Biology 38:293-382.

Giocomo, J.J., M. Gustafson, J.N. Duberstein, and C. Boyd. 2012. The role of Joint Ventures in bridging the gap between research and management. In Wildlife Science: Connecting Research with Management, edited by J.P. Sands, S.J. DeMaso, M.J. Schnupp, and L.A. Brennan. Boca Raton, FL: CRC Press, pp. 239-251.

Gleason, M., S. McGovern, M. Miller-Henson, J. Ugoretz, E. Fox, M. Merrifield, W. McClintock, P. Serpa, and K. Hoffman. 2010. Science-based and stakeholder-driven marine protected area network planning: A successful case study from north-central California. Ocean and Coastal Management 53:52-68.

Goldstein, B.E., and W.H. Butler. 2012. Collaborating for transformative resilience: Shared identity in the U.S. Fire Learning Network. In Collaborative Resilience: Moving Through Crisis to Opportunity, edited by B.E. Goldstein. Cambridge, MA: MIT Press, p. 339.

González, J.A., C. Montes, J. Rodríguez, and W. Tapia. 2008. Rethinking the Galapagos Islands as a complex social-ecological system: Implications for conservation and management. Ecology and Society 13(2):13. Available at http://www.ecologyandsociety.org/vol13/iss2/art13 (accessed October 5, 2016).

Gore, A. 1993. From Red Tape to Results: Creating a Government That Works Better and Costs Less. Report of the National Performance Review, pp. 52-53.

Groves, C.R. 2003. Drafting a Conservation Blueprint: A Practitioner's Guide to Planning for Biodiversity. Washington, DC: Island Press.

Groves, C.R., and E.T. Game. 2015. Conservation Planning: Informed Decisions for a Healthier Planet. Englewood, CO: Roberts and Company.

Groves, C.R., D.B. Jensen, L.L. Valutis, K.H. Redford, M.L. Shaffer, J.M. Scott, J.V. Baumgartner, J.V. Higgins, M.W. Beck, and M.G. Anderson. 2002. Planning for biodiversity conservation: Putting conservation science into practice. BioScience 52(6):499-512.

Hicks, C.C., J.E. Cinner, N. Stoeckl, and T.R. McClanahan. 2015. Linking ecosystem services and human-values theory. Conservation Biology 29(5):1471-1480.

Hughson, D.L., D.E. Busch, S. Davis, S.P. Finn, S. Caicco, and P.S.J. Verburg. 2011. Natural Resource Mitigation, Adaptation and Research Needs Related to Climate Change in the Great Basin and Mojave Desert: Workshop Summary. U.S. Geological Survey Scientific Investigations Report 2011-5103, 34 pp.

Hunter, M.L., G.L. Jacobson, Jr., and T. Webb III. 1988. Paleoecology and the coarse-filter approach to maintaining biological diversity. Conservation Biology 2:375-385.

Imperial, M., E. Johnston, M. Pruett-Jones, K. Leong, and J. Thomsen. In press. Sustaining the useful life of network governance: Life-cycles and developmental challenges. Frontiers in Ecology and the Environment.

Intergovernmental Oceanographic Commission. 2009. Marine Spatial Planning: A Step-by-Step Approach Toward Ecosystem-Based Management, edited by R. Dahl. Paris: United Nations Educational, Scientific, and Cultural Organization.

Jacobson, C., and A.L. Robertson. 2012. Landscape Conservation Cooperatives: Bridging entities to facilitate adaptive co-governance of social-ecological systems. Human Dimensions of Wildlife 17:333-343.

Kareiva. P., H. Tallis, T.H. Ricketts, G.C. Daily, and S. Polasky. 2011. Natural Capital. Theory and Practice of Mapping Ecosystem Services. Oxford, UK: Oxford University Press.

Kati, V., T. Hovardas, M. Dieterich, P.L. Ibisch, B. Mihok, and N. Selva. 2015. The challenge of implementing the European network of protected areas Natura 2000. Conservation Biology 29(1):260-270.

Keener, V.W., J.J. Marra, M. Finucane, D. Spooner, and M.H. Smith, eds. 2012. Climate Change and Pacific Islands: Indicators and Impacts. Report for the 2012 Pacific Islands Regional Climate Assessment. Washington, DC: Island Press.

Keeney, R.L. 1992. Value-Focused Thinking: A Path to Creative Decision Making. Cambridge, MA: Harvard University Press.

Knick, S.T., S.E. Hanser, and K.L. Preston. 2013. Modeling ecological minimum requirements for distribution of greater sage-grouse leks: Implications for population connectivity across their western range, U.S.A. Ecology and Evolution 3:1539-1551.

Knight, A.T., R.M. Cowling, M. Rouget, A. Balmford, A.T. Lombard, and B.M. Campbell. 2008. Knowing but not doing: Selecting priority conservation areas and the research-implementation gap. Conservation Biology 22(3):610-617.

Landau, M. 1969. Redundancy, rationality, and the problem of duplication and overlap. Public Administration Review 29(4):346-358.

Lauber, T.B., R.C. Stedman, D.J. Decker, and B.A. Knuth. 2011. Linking knowledge to action in collaborative conservation. Conservation Biology 25(6):1186-1194.

Laven, D., C. Ventriss, R. Manning, and N. Mitchell. 2010. Evaluating U.S. National Heritage Areas: Theory, methods, and application. Environmental Management 46:195-212.

Lawler, J.J., D.J. Lewis, E. Nelson, A.J. Plantinga, S. Polasky, J.C. Whitney, D.P. Helmers, S. Martinuzzi, D. Pennington, and V.C. Radeloff. 2014. Projected land-use change impacts on ecosystem services in the United States. Proceedings of the National Academy of Sciences of the United States of America 111(20):7492-7497.

LCC (Landscape Conservation Cooperatives). 2014. LCC Network Strategic Plan. Available at http://lccnetwork.org/strategic-plan (accessed October 5, 2016).

LCC Network. 2011. Landscape Conservation Cooperatives: Operational Memo for Addressing Requests for Consideration of LCC Geographic Boundary Changes. December 6.

LCC Network. 2014. Mississippi River Basin/Gulf Hypoxia Conservation Design Briefing Memo. Available at https://griffingroups.com/file/view/164063/lccs-mrb-gulf-hypoxia-briefing-memo (accessed October 5, 2016).

LCC Network, LCC Science Coordinators Team. 2015. LCC Network Conservation Science Plan Version 1.0. Available at http://lccnetwork.org/resource/science-plan (accessed April 12, 2016).

Leopold, A. 1947. Draft foreword to A Sand County Almanac. In Companion to a Sand County Almanac: Interpretive and Critical Essays, edited by J. Baird Callicott. Madison: University of Wisconsin Press.

Leu, M., S.E. Hanser, and S.T. Knick. 2008. The human footprint in the west: A large-scale analysis of anthropogenic impacts. Ecological Applications 18(5):1119-1139.

Lind, E.A., and T.R. Tyler. 1988. The Social Psychology of Procedural Justice. New York: Plenum.

Lindblom, C. 1968. The Policy-Making Process. Englewood Cliffs, NJ: Prentice Hall.

Mace, G.M. 2014. Whose conservation? Changes in the perception and goals of nature conservation require a solid scientific basis. Nature 345:1558-1560.

Margerum, R.D. 2008. A typology of collaboration efforts in environmental management. Environmental Management 41:487-500.

Marsh, G.P. 1864. Man and Nature: Or, Physical Geography as Modified by Human Action. New York: Charles Scribner.

Marsh, W.M. 2010. Landscape Planning: Environmental Applications. Somerset, NJ: John Wiley and Sons.

Martin-Williams, S. 2007. National Heritage Areas: Developing and Specifying a Model of Inter-organizational Domain Development and Exploring the Role of the National Park Service as a Federal Partner. Morgantown: West Virginia University.

Mascia, M.B., S. Pailler, M.L. Thieme, A. Rowe, M.C. Bottrill, F. Danielsen, J. Geldmann, R. Naidoo, R.S. Pullin, and N.D. Burgess. 2014. Commonalities and complementarities among approaches to conservation monitoring and evaluation. Biological Conservation 169:258-267.

Mayer, K.E., P. Anderson, J. Chambers, C. Boyd, T. Christiansen, D. Davis, S. Espinosa, D. Havlina, M. Ielmini, D. Kemner, L. Kurth, J. Maestas, B. Mealor, T. Milesneck, L. Niell, M. Pellant, D. Pyke, J. Tague, and J. Vernon. 2013. Wildfire and Invasive Species in the West: Challenges That Hinder Current and Future Management and Protection of the Sagebrush Steppe Ecosystem—A Gap Report. Cheyenne, WY: Western Association of Fish and Wildlife Agencies.

McKinney, M., and S. Johnson. 2009. Working Across Boundaries: People, Nature, and Regions. Cambridge, MA: Lincoln Institute of Land Policy.

McKinney, M., L. Scarlett, and D. Kemmis. 2010. Large landscape conservation: A strategic framework for policy and action. Cambridge, MA: Lincoln Institute of Land Policy.

MEA (Millennium Ecosystem Assessment). 2005. Summary for decision makers. In Ecosystems and Human Well-Being: Synthesis. Washington, DC: Island Press, pp. 1-24.

Meadow, A., D. Ferguson, Z. Guido, A. Horangic, G. Owen, and T. Wall. 2015. Moving toward the deliberate co-production of climate science knowledge. Weather, Climate, and Society 7(2):179-191.

Millard, M.J., C.A. Czarnecki, J.M. Morton, L.A. Brandt, J.S. Briggs, F.S. Shipley, R. Sayre, P.J. Sponholtz, D. Perkins, D.G. Simpkins, and J. Taylor. 2012. A national geographic framework for guiding conservation on a landscape scale. Journal of Fish and Wildlife Management 3(1):175-183.

Milner-Gulland, E.J., J.A. McGregor, M. Agarwala, G. Atkinson, P. Gevan, T. Clements, T. Daw, K. Homewood, N. Kumpel, J. Lewis, S. Mourato, B. Palmer, M. Redshaw, J.M. Rowcliffe, S. Suon, G. Wallace, H. Washington, and D. Wilkie. 2014. Accounting for the impact of conservation on human well-being. Conservation Biology 28:1160-1166.

Mitchell, R. K., H. J. VanBuren III, M. Greenwood, and R. E. Freeman. 2015. Stakeholder inclusion and accounting for stakeholders. Prepublication version. Journal of Management Studies 52(7):851-877.

Moffett, A., and S. Sarkar. 2006. Incorporating multiple criteria into the design of conservation area networks: A minireview with recommendations. Diversity and Distributions 12:125-137.

MRGMWNTF (Mississippi River/Gulf of Mexico Watershed Nutrient Task Force). 2008. Gulf Hypoxia Action Plan 2008 for Reducing, Mitigating, and Controlling Hypoxia in the Northern Gulf of Mexico and Improving Water Quality in the Mississippi River Basin. Washington, DC: MRGMWNTF.

MRGMWNTF. 2013. Reassessment 2013: Progress Made Since 2008. Available at http://water.epa.gov/type/watersheds/named/msbasin/upload/hypoxia_reassessment_508.pdf (accessed November 20, 2015).

National Fish Habitat Action Plan. 2012. Available at http://fishhabitat.org/sites/default/files/www/NFHP_AP_Final_0.pdf (accessed October 5, 2016).

National Technical Assistance Team. 2008. Strategic Habitat Conservation Handbook: A Guide to Implementing the Technical Elements of Strategic Habitat Conservation, Version 1.0. Washington, DC: U.S. Fish and Wildlife Service.

NAWMP (North American Waterfowl Management Plan) Committee. 2012. North American Waterfowl Management Plan 2012: People Conserving Waterfowl and Wetlands. Available at http://www.dgif.virginia.gov/wildlife/waterfowl/north-american-waterfowl-management-plan-2012.pdf (accessed September 8, 2016).

Noss, R.F. 1987. From plant communities to landscapes in conservation inventories: A look at The Nature Conservancy (USA). Biological Conservation 41:11-37.

Noss, R.F. 2002. Context matters: Considerations for large-scale conservation. Conservation Biology in Practice 3:10-19.

NPS (National Park Service). 2011. A Call for Action: Preparing for a Second Century of Stewardship and Engagement. NPS.

NRC (National Research Council). 2003. Does Water Flow Influence Everglades Landscape Patterns? Washington, DC: The National Academies Press.

NRC. 2004. Valuing Ecosystem Services: Toward Better Environmental Decision-Making. Washington, DC: The National Academies Press.

NRC. 2005a. Assessing and Managing the Ecological Impact of Paved Roads. Washington, DC: The National Academies Press.

NRC. 2005b. Re-Engineering Water Storage in the Everglades: Risks and Opportunities. Washington, DC: The National Academies Press.

NRC 2005c. Thinking Strategically: The Appropriate Use of Metrics for the Climate Change Science Program. Washington, DC: The National Academies Press.

NRC. 2007. Status of Pollinators in North America. Washington, DC: The National Academies Press.

NRC. 2010. Adapting to the Impacts of Climate Change. Washington, DC: The National Academies Press.

NRC. 2011. A Review of the Use of Science and Adaptive Management in California's Draft Bay Delta Conservation Plan. Washington, DC: The National Academies Press.

NRC. 2012a. Improving Water Quality in the Mississippi River Basin and the Northern Gulf of Mexico: Strategies and Priorities. Washington, DC: The National Academies Press.

NRC. 2012b. Sustainable Water and Environmental Management in the California Bay-Delta. Washington, DC: The National Academies Press.

NRC. 2014a. Mississippi River Water Quality and Interstate Collaboration: Summary of a Workshop. Washington, DC: The National Academies Press.

NRC. 2014b. Progress Toward Restoring the Everglades: The Fifth Biennial Review. Washington, DC: The National Academies Press.

O'Connell, A.J. 2006. The architecture of smart intelligence: Structuring and overseeing agencies in the post-9/11 world. California Law Review 94:1655-1744.

Office of the Science Advisor. 2010. LCC Information Bulletin No. 1: Form and Function. U.S. Fish and Wildlife Service. Available at http://www.fws.gov/southeast/LCC/pdf/LCCBulletin1FormandFunction01162010.pdf (accessed September 8, 2016).

Omernik, J.M. 1995. Ecoregions: A spatial framework for environmental management. In Biological Assessment and Criteria: Tools for Water Resource Planning and Decision Making, edited by W.S. Davis and T.P. Simon. Boca Raton, FL: Lewis, pp. 49-62.

Omernik, J.M. 2004. Perspectives on the nature and definition of ecological regions. Environmental Management 34(Suppl. 1):S27-S38.

Ostrom, E. 1990. Governing the Commons: The Evolution of Institutions for Collective Action. Cambridge, UK: Cambridge University Press.

Patrizi, P., M. Lempa, E. Wilson, and L. Albright. 2009. Lessons Learned from the Pennsylvania Department of Conservation and Natural Resources Conservation Landscape Initiatives, OMG Center for Collaborative Learning, December 2009. Available at http://www.dcnr.state.pa.us/cs/groups/public/documents/document/d_001846.pdf (accessed October 5, 2016).

Pauley, T.K. 1993. Amphibians and reptiles of the upland forests. In Upland Forests of West Virginia, edited by S.L. Stephenson. Parsons, WV: McClain Printing Company, pp. 179-196.

Peterson, B., J. Montambault, and M. Koopman. 2014. The potential for double-loop learning to enable landscape conservation efforts. Environmental Management 54(4):782-794.

Pimm, S., C.N. Jenkins, R. Abell, T.M. Brooks, J. Gittleman, L.N. Joppa, P.H. Raven, C.M. Roberts, and J.O. Sexton. 2014. The biodiversity of species and their rates of extinction, distribution, and protection. Science 344 (6187):1246752.

Poiani, K.A., B.D. Richter, M.G. Anderson, and H.E. Richter. 2000. Biodiversity conservation at multiple scales: Functional sites, landscapes, and networks. Bioscience 50(2):133-146.

Powell, W.W. 1990. Neither market nor hierarchy: Network forms of organization. Research in Organizational Behavior 12:295-336.

Pressey, R.L., and M.C. Bottrill. 2009. Approaches to landscape- and seascape-scale conservation planning: Convergence, contrasts and challenges. Oryx 43(4):464-475.

Pulwarty, R.S., C. Simpson, and C.R. Nierenberg. 2009. The Regional Integrated Sciences and Assessments (RISA) program: Crafting effective assessments for the long haul. In Integrated Regional Assessment: Challenges and Case Studies, edited by C.G. Knight and J. Jäger. Cambridge, UK: Cambridge University Press, pp. 367-393.

Redford, K.H., P. Coppolillo, E.W. Sanderson, G.A.B. Da Fonseca, E. Dinerstein, C. Groves, G. Mace, S. Maginnis, R.A. Mittermeier, R. Noss, D. Olson, L.G. Robinson, A. Vedder, and M. Wright. 2003. Mapping the conservation landscape. Conservation Biology 17(1):116-131.

Reyers, B., D.J. Roux, R.M. Cowling, A.E. Ginsburg, J.L. Nel, and P. O'Farrell. 2010. Conservation planning as a transdisciplinary process. Conservation Biology 24(4):957-965.

Sayre, N.F. 2005. Working Wilderness: The Malpai Borderlands Group and the Future of the Western Range. Tucson, AZ: Rio Nuevo Press.

Sayer, J., T. Sunderland, J. Ghazoul, J.-L. Pfund, D. Sheil, E. Meijgaard, M. Venter, A.K. Boedhihartono, M. Day, C. Garcia, C. van Oosten, and L.E. Buck. 2013. Ten principles for a landscape approach to reconciling agriculture, conservation, and other competing land uses. Proceedings of the National Academy of Sciences of the United States of America 110(21):8349-8356.

Scarlett, L. 2013. Collaborative adaptive management: Challenges and opportunities. Ecology and Society 18(3):26.

Scherr, S.J., and J.A. McNeely. 2008. Biodiversity conservation and agricultural sustainability: Towards a new paradigm of "ecoagriculture" landscapes. Philosophical Transactions of the Royal Society of London B: Biological Sciences 363(1491):477-494.

Schwartz, M. 1999. Choosing the appropriate scale of reserves for conservation. Annual Review of Ecology and Systematics 30:83-108.

SIAS (Science Investment and Accountability Schedule). 2013. SIAS 2.0: Background and Instructions. Available at https://my.usgs.gov/confluence/download/attachments/436174920/SIAS_2.0_background-and-instructions_v20131206_FINAL.pdf?api=v2 (accessed October 5, 2016).

Slocombe, D.S. 1993. Implementing ecosystem-based management. BioScience 43:612-622.

Soulé, M., and J. Terborgh. 1999. Conserving nature at regional and continental scales: A scientific program for North America. Bioscience 49(10):809-817.

South Atlantic LCC. 2015. Conservation Blueprint 2.0. Available at http://salcc.databasin.org/maps/a46404d870df478f871e1af23d8da539 (accessed October 5, 2016).

Terwilliger Consulting. 2013. Taking Action Together: Northeast Regional Synthesis for State Wildlife Actions Plans. Prepared for the Northeast Fish and Wildlife Diversity Committee.

Theobald, D. 2014. Development and applications of a comprehensive land use classification and map for the U.S. PLoS ONE 9(4):e94628.

Thomson, A.M. and J.L. Perry, 2010. Collaboration processes: Inside the black box. In: Public Administration: Concepts and Cases, edited by R.J. Stillman II. Boston: Wadsworth CENGAGE Learning.

Thomson, A.M., J.L. Perry, and T.K. Miller. 2007. Conceptualizing and measuring collaboration. Journal of Public Administration and Theory 19(1):23-56. Available at http://www.indiana.edu/~jlpweb/papers/Conceptualizing%20and%20Measuring%20Collab_Thomson_Perry_Miller_JPART_Dec%202007.pdf (accessed October 5, 2016).

Turner, M. 1989. Landscape ecology: The effect of pattern on process. Annual Review of Ecology and Systematics 20:171-197.

USGS (U.S. Geological Survey). 2008. The Gulf of Mexico Hypoxic Zone. Available at http://toxics.usgs.gov/hypoxia/hypoxic_zone.html (accessed October 5, 2016).

Walters, C.J., and C.S. Holling. 1990. Large-scale management experiments and learning by doing. Ecology 71(6):2060-2068.

Wholey, J.S. 1996. Formative and summative evaluation: Related issues in performance measurement. American Journal of Evaluation 17(2):145-149.

Wood, D., and B. Gray. 1991. Towards a comprehensive theory of collaboration. Journal of Applied Behavioral Science 27:139-162.

Yellowstone to Yukon Conservation Initiative. 2014. The Yellowstone to Yukon Vision Progress and Possibilities: Celebrating 20 Years of the Y2Y Vision. Yellowstone to Yukon Conservation Initiative.

Young, J.C., A. Jordan, K.R. Searle, A. Butler, D.S. Chapman, P. Simmons, and A.D. Watt. 2013. Does stakeholder involvement really benefit biodiversity conservation? Biological Conservation 158:359-370.

Appendix A

Greater Sage-Grouse: A Collaborative Conservation Effort

INTRODUCTION

The greater sage-grouse case study illustrates the importance of partnerships between private landowners and state and federal agencies; how challenging it can be for a Landscape Conservation Cooperative (LCC) to integrate into an existing landscape-scale issue; how the scale at which conservation action needs to take place can vary; and how difficult it is to attribute credit to any single entity in such a collaborative effort. The discussion below illustrates how a landscape approach can catalyze and facilitate conservation that averts the listing of a species. The narrative demonstrates the benefits of the landscape approach for addressing emerging conservation priorities and how voluntary conservation actions can help to avoid a species decline to the point where it requires listing. Maintaining healthy population levels to avoid the listing "trigger" benefits the target species but also benefits a broad range of stakeholders such as public land managers, private landowners, ranching interests, and oil, gas, and mineral developers who could potentially lose some flexibility in how they manage their resources if a species becomes listed.

Although LCCs did not play a central role in this case, they were established to identify such priority species or habitats and to facilitate collaboration to yield the type of collaborative conservation effort that is described below. It became clear when reviewing the LCC activities related to sage-grouse that LCCs joined an already very mature partnership and the LCC extent of contribution is not obvious. In almost every activity listed above, the LCC participated as a stakeholder in an initiative but seldom appeared to be the catalyst for the initiative. Notwithstanding, their funds and emphasis on landscape problems have been helpful in supporting some very important work on greater sage-grouse.

The LCCs also have contributed to the greater sage-grouse conservation effort in more subtle ways. For example, the Western Association of Fish and Wildlife Agencies (WAFWA), the state of Nevada, and several federal agencies combined resources to prepare a strategic approach based on resistance and resilience concepts for conservation of sagebrush ecosystems and greater sage-grouse (Chambers et al., 2014). While the direct funding for this effort came from state and federal agencies, the Great Basin LCC provided staff support for a GIS and modeling expert who played a key role in the creation of a geospatial tool required to implement the Fire and Invasive Species Team protocol described by Chambers et al. (2014).

However, it also illustrates that the LCC program has played an important role in several initiatives, especially those that span multiple jurisdictions, interactions among multiple species, and issues that play out over long timescales (i.e., those aspects that are most difficult for other agencies to address).

The sage-grouse example also highlights the fact that the existing LCC boundaries will not always match the scale of landscape-scale issues. This is not surprising as no administrative boundary will perfectly match the distribution of species or landscapes. Even the administration of the LCCs results in some difficulty in coordinating landscape-scale research for a species such as the greater sage-grouse. For example, the Inter-LCC Greater Sage-Grouse initiative was administered out of Region 6, even though the region has administrative responsibility for only two of the four LCCs containing habitat for the species. LCC boundaries may constrain conversations. Bird Conservation Regions with modifications based on terrestrial and freshwater ecoregions (Gallant, 1989; Omernik, 1995, 2004; Abell et al., 2000) are the basis for the LCC boundaries, but many of the landscape-scale issues are not captured within these boundaries.

It is important to note that this sage-grouse conservation effort was initiated long before the establishment of the LCCs. The case study intends to illustrate how voluntary conservation partnerships can avert the decline of a target species and avoid listing under the Endangered Species Act (ESA). Keeping the greater sage-grouse off the endangered species list is beneficial not only to the survival of the spe-

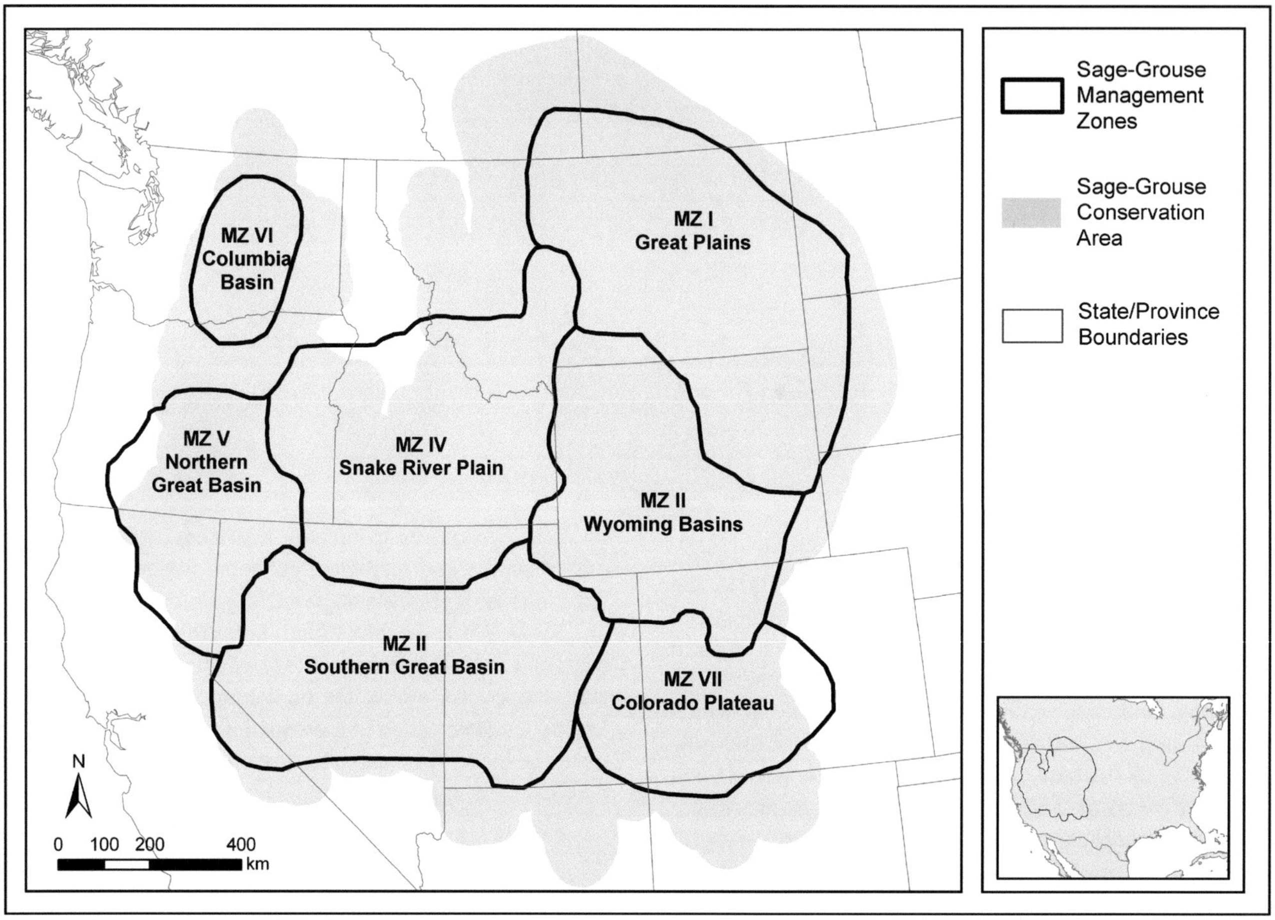

FIGURE A.1 Greater sage-grouse populations and sage-grouse management zones in western North America.
SOURCE: Garton et al., 2011.

cies in question, but also to the broad range of stakeholders mentioned above. The following case study demonstrates the range of actions and partnerships that are contributing to this conservation effort.

The Target Species Range and Ecology

The greater sage-grouse is endemic to the sagebrush steppe landscape in 11 states in the United States and in two Canadian provinces (see Figure A.1). The greater sage-grouse is protected by state law throughout its range and managed as an upland gamebird by state wildlife agencies. The species is experiencing range-wide population declines due to agricultural development, large-scale range improvements (sagebrush control), urban and exurban development, large wildfires, invasion of exotic plants, and more recently, energy development.

Conservations Efforts Led by State Agencies

Western states have a longstanding practice for addressing population and habitat conservation across state lines through a range of cross-state partnerships. Recognizing the decline of the sage-grouse populations, members of WAFWA signed a "Memorandum of Understanding [MOU] Among Members of the Western Association of Fish and Wildlife Agencies for the Conservation and Management of Sage-Grouse in North America."[1] The MOU was expanded in 2000 to include federal agencies (U.S. Forest Service [USFS], Bureau of Land Management [BLM], U.S. Fish and Wildlife Service [FWS], U.S. Geological Survey [USGS], Natural Resources Conservation Service [NRCS], and Farm Service Agency).

In addition to collaborating with the FWS through federal initiatives, many of the 11 states where greater sage-grouse habitat occurs have identified important areas needing special consideration in land-use decisions. California and Nevada have been working with the BLM, USFS, NRCS, and the FWS to conserve the bi-state sage-grouse population, considered a distinct population by the FWS for more than a decade. In March 2012, the bi-state Executive Oversight

[1] See http://www.wafwa.org/initiatives/sage_steppe/Sage-Grouse ConservationImplementationMOU.pdf.

Committee for Conservation of Greater Sage-Grouse published the *Bi-state Action Plan—Past, Present, and Future Actions for Conservation of the Greater Sage-Grouse Bi-state Distinct Population Segment* (Bi-State Technical Advisory Committee Nevada and California, 2013). The action plan summarizes the steps that have been taken to conserve the bi-state population of greater sage-grouse and identified objectives and strategies guiding future conservation efforts. The states of Montana, Oregon, Utah, and Wyoming have identified core habitat areas within their states that are considered essential for the maintenance of sage-grouse in those states.

In Wyoming, for example, the governor issued the Sage-Grouse Executive Order (SGEO) through the regulatory authority of state agencies and collaboration with industry, federal land management agencies, and local sage-grouse working groups. The SGEO provides for coordination of greater sage-grouse conservation efforts among all stakeholders statewide and an evaluation of projects within greater sage-grouse core areas, and prohibits state agencies in most cases from taking actions leading to the loss of core habitat.

The governor of Idaho established the Idaho Sage-Grouse Task Force in 2012 to "prevent the need for the federal protection under ESA."[2] The Task Force developed through broad stakeholder engagement an alternative management plan to listing the species under ESA.[3] It identified three distinct management areas, ranging from management with a "restrictive approach" to management areas that would allow "multiuse activities."

The Western Governors' Association (WGA) and WAFWA play an important role in coordinating greater sage-grouse conservation initiatives among states and federal agencies. In their *Sage Grouse Inventory 2014 Conservation Initiatives*, the WGA summarized the activities taking place to conserve the sage-grouse in all 11 states within the species range. WAFWA focused on educating its membership about the species and different measures needed for its conservation. In their "Gap Report," Mayer et al. (2013) recognized that wildfire and the subsequent spread of invasive plants continue to play a huge role in the loss of sagebrush steppe habitat, particularly in the western portion of the greater sage-grouse range. Their report summarized the policy, fiscal, and science challenges that land managers encounter related to the control and reduction of the "invasive plant/fire complex," in relation to greater sage-grouse conservation. WAFWA also provides a clearinghouse for sources of information relevant to important initiatives related to greater sage-grouse and a vehicle for states and federal agencies to pool funds for large-scale projects.

[2] See https://fishandgame.idaho.gov/public/wildlife/?getPage=310.

[3] See https://fishandgame.idaho.gov/public/wildlife/SGtaskForce/alternative.pdf.

The U.S. Fish and Wildlife Service's Listing Process

The FWS was petitioned to list the greater sage-grouse under ESA in 2002 and again in 2003. On January 12, 2005, the agency ruled that a listing as threatened or endangered was not warranted. In 2008, the FWS announced a status review and requested new information for the species in response to new listing petitions. The FWS reached the conclusion in 2010 that listing of the greater sage-grouse was warranted due to habitat loss and fragmentation and inadequacy of regulatory mechanisms that govern habitat loss and fragmentation. The agency identified fragmentation of sagebrush landscapes as one of the primary causes of the decline of greater sage-grouse populations. The direct loss of habitat on a large scale due to invasive species following wildfires, agricultural conversion of sagebrush landscapes to grasslands, and poor grazing practices is exacerbated by displacement from otherwise suitable habitat and functional habitat loss due to energy and other infrastructure developments. Nevertheless, the FWS determined that the listing was precluded because of higher-priority listing actions. On September 30, 2015, the "U.S. Fish and Wildlife Service (FWS) . . . conclude[d] that the charismatic rangeland bird does not warrant protection under the Endangered Species Act (ESA)."[4] In this press release, the secretary of the U.S. Department of the Interior credits the voluntary conservation efforts and the partnerships between federal and state agencies as well as private landowners and conservation groups for contributing to this historic decision.

Conservations Efforts Supported by Federal Agencies

In 2012, the FWS convened the Conservation Objectives Team to initiate a collaborative approach with the states to develop range-wide conservation objectives to both inform the 2015 listing decision and to provide guidance to measures being taken to conserve the greater sage-grouse by individual states and agencies. The FWS has also developed a number of partnerships throughout the 11-state region occupied by the species in an effort to accomplish as many greater sage-grouse conservation projects as possible and to ensure as much protection for their habitat as possible prior to the 2015 listing deadline.

A number of the federal partners have created Candidate Conservation Agreements (CCAs) with private landowners—CCAs with Assurances (CCAA)—for the greater sage-grouse under Section 10(a)(1)(A) of ESA.[5] Both are

[4] See https://www.doi.gov/pressreleases/historic-conservation-campaign-protects-greater-sage-grouse.

[5] CCAs are formal, voluntary agreements between the FWS and one or more parties to address the conservation needs of one or more candidate species or species likely to become candidates in the near future. Participants voluntarily commit to implement specific actions designed to remove or reduce threats to the covered species, so that listing may not be necessary (http://www.fws.gov/endangered/esa-library/pdf/CCAs.pdf). Because a permit is not issued for a CCA, there are no assurances that conservation

conservation agreements between the FWS and other parties to initiate voluntary conservation actions to avoid listing of candidate species. Examples of such conservation agreements include the following:

1. The BLM entered into a number of CCAs with the FWS for rangeland management in Oregon and Wyoming.
2. The FWS recently announced a complementary CCAA in Oregon allowing landowners in all eight eastern and central Oregon counties with greater sage-grouse habitat to enroll in the voluntary program.

Another approach for conservation of greater sage-grouse habitat is led by the U.S. Department of Agriculture's NRCS, which created the Sage Grouse Initiative that includes the NRCS, ranchers, state and federal agencies, universities, nonprofit groups, and private businesses. In the past 5 years, the Sage Grouse Initiative has leveraged the NRCS investment with additional funds from partners and landowners to a total investment of $424.5 million. The Sage Grouse Initiative has been targeted toward the most important regions and aids ranchers with NRCS technical and financial assistance and in getting NRCS conservation practices on the ground. Efforts range from establishing conservation easements that prevent subdivision of large and intact working ranches to improving and restoring habitat through removal of invasive trees. Across the range, conservation easements have increased 18-fold through the Sage Grouse Initiative, protecting 451,884 acres.

Privately Led Efforts

Efforts to conserve greater sage-grouse habitat are not limited to state and federal agencies. Industry and private landowners are also developing means of conserving the greater sage-grouse. In April 2015, the Secretary of the Interior announced an agreement with Barrick Gold of North America to create a greater sage-grouse conservation bank that protects important habitat for the species on lands controlled by the mining company as mitigation for impacts caused as the company proceeds with their gold mining activity.

In March 2015, the Secretary of the Interior announced the approval of the privately held Sweetwater River Conservancy Conservation Bank for the greater sage-grouse on more than 55,000 deeded acres in central Wyoming. Not only was this the first conservation bank for greater sage-grouse, it is the largest habitat conservation bank in the United States. The bank establishes habitat credits, based on extensive data on greater sage-grouse habitat use and habitat preference and the identification of functional population segments over a wide area including private, state, and federal lands. The credits can be used to offset impacts to greater sage-grouse within the bank's service area, essentially the range of the species within the state. The private lands are protected through a perpetual easement and managed in accordance with the Conservation Bank Agreement approved by the FWS. The revenue for protection and enhancement of this habitat is generated through the sale of the habitat credits. According to the FWS the bank has the potential to expand to 700,000 acres on lands owned by the Sweetwater River Conservancy.

measures will not change should listing occur. In contrast, a CCAA is a voluntary agreement between the FWS and participating private property owners with a permit issued by the agency containing assurances that if the private property owners engage in certain conservation actions for species included in the agreement, they will not be required to implement additional conservation measures beyond those in the CCAA if the species is listed in the future. Also, in the case of a CCAA, additional land, water, or resource use limitations will not be imposed on them should the species become listed in the future, unless the landowners consent to such changes necessary (http://www.fws.gov/endangered/esa-library/pdf/CCAs.pdf).

LCC ROLE IN CONSERVATION OF THE GREATER SAGE-GROUSE

The historic and current range of the greater sage-grouse includes portions of four LCCs. Although the greater sage-grouse conservation issue predates the development of the LCCs, the LCC Network has contributed $2.25 million and staff support to 27 greater sage-grouse projects. Much of the LCC money and staff were leveraged, allowing stakeholders to tackle much larger problems than would have been possible individually. For example, the Inter-LCC Collaborative Sage-Grouse Project received $500,000 from the FWS, which they leveraged for $941,000 of other funding for specific projects with range-wide application and demonstrating their collaborative nature. The following are examples of projects that were funded to contribute to existing sage-grouse conservation efforts mentioned above.

Great Northern LCC

In their draft Science Plan, the Great Northern LCC (GNLCC) lists the greater sage-grouse and its habitat, sagebrush/grassland, as conservation targets and the greater sage-grouse as an umbrella species for the sagebrush steppe landscape. The GNLCC has contributed more than $750,000 to eight ongoing or completed projects either directly or indirectly related to the greater sage-grouse. Examples of what these studies contributed follow:

- Created a habitat model for greater sage-grouse in the Columbia Plateau Ecoregion; tested connectivity model predictions for greater sage-grouse and focal species tied to sage-steppe ecotypes (black- and white-tailed jackrabbits); integrated model testing results in an adaptive management framework to inform conservation action within the area of the GNLCC, and communicated to shareholders.
- Provided high-resolution connectivity maps for greater sage-grouse in the Great Northern landscape using state-of-the-art genomics. The project resulted in a manu-

script prepared for publication and allowed the identification of genes that code for disease resistance and adaptive selection, such that consideration of adaptive differences can be used in management of subpopulations.

- Monitored for 5 years vegetation, fuels, wildlife, insects, and weather at 10 Sagebrush Steppe Treatment Evaluation Project (SageSTEP) sites, all of which were treated to reduce either juniper encroachment (woodland sites) or cheatgrass invasion (sagebrush/cheatgrass sites).
- Identified greater sage-grouse populations at risk of extinction within the GNLCC based on their relative isolation from neighboring populations and core regions of the greater sage-grouse. These results benefited management agencies by focusing regional conservation and land management options in regions likely to sustain long-term sagebrush ecosystems (Knick et al., 2013).

Great Basin LCC

For 2013 and 2014, the Great Basin LCC identified short-term science priorities to guide research (Hughson et al., 2011). The Great Basin LCC is preparing a longer-term plan to guide research for the next 5 years. The Great Basin LCC has contributed more than $825,000 in the funding of 10 projects directly or indirectly related to the greater sage-grouse. Examples of these projects include the following:

- Assessment of impacts of feral horses and livestock grazing on sage-grouse habitats: long-term trends in greater sage-grouse demography and habitats on the Sheldon-Hart Mountain. This project takes advantage of historical patterns of grazing by both feral horses and livestock and new data to assess greater sage-grouse population dynamics and habitats under all combinations of grazing by nonnative ungulates.
- Forecasting changes in sagebrush distribution and abundance under climate change: integration of spatial, temporal, and mechanistic models. The goal of this project is to forecast the effect of climate change on the distribution and abundance of big sagebrush in order to inform conservation planning, and sage-grouse management in particular, across the Intermountain West. The novelty of the work will be the synthesis of models based on spatial, temporal, and mechanistic relationships between climate and sagebrush cover.
- Strategic high-resolution wetland mapping in greater sage-grouse biologically significant areas of Nevada. This effort is a direct result of the Great Basin LCC-led Central Basin & Range Rapid Ecoregional Assessment (CBR REA) Challenges and Opportunities Report, which identified a paucity of available wetland and springs data layers for the CBR REA area. This project will provide wetland mapping at high resolution (1:24,000) for 13 million acres of greater sage-grouse biologically significant areas within Nevada.

Plains and Prairie Potholes LCC

We found no formal plan for research for the Plains and Prairie Potholes LCC (PPLCC) or any specific research objectives related to the greater sage-grouse. However, the LCC provided support for three projects directly or indirectly related to the greater sage-grouse, including

- An evaluation of the impact of conservation-oriented, rest-rotation livestock grazing and climate changes on migratory bird species associated with sagebrush habitat to better inform grazing management practices. While this project was not directly related to the greater sage-grouse, rest-rotation grazing management is likely to enhance important components of sagebrush, shrubland, and grassland habitat for a wide range of species including greater sage-grouse.
- An investigation of the construction and operational effects of wind energy development on greater sage-grouse through the study of survival, movements, habitat use, and lek dynamics, on a 1,000-turbine, 2,000- to 3,000-megawatt wind facility in Carbon County, Wyoming, using a before-after control-impact design. Another study was successful in addressing the study objective of evaluating the impact of wind energy development on greater sage-grouse.
- The development of a rapid assessment method for wildlife issues at potential wind energy sites.

Southern Rockies LCC

No research plan was found for the Southern Rockies LCC (SRLCC), the fourth LCC including a portion of the current range of the greater sage-grouse. The list of ongoing or completed projects on the SRLCC website included two projects related to the greater sage-grouse: $250,000 funding contributed to the Western States Crucial Habitat Assessment Tools; and contributions to the development of a Regional Model for Building Resilience to Climate Change: Development and Demonstration in Colorado for the Gunnison sage-grouse.

Inter-LCC Sage-Grouse Initiative

Because of the nature of the boundaries of the states and the LCCs, it is difficult to coordinate research and management on a wide-ranging species such as the greater sage-grouse. While federal and state governments have worked tirelessly at trying to collaborate on transboundary initiatives, the four LCCs have been focused on smaller-scale projects typically on specific issues within each LCC boundary. And in the case of the SRLCC, the focus has been heavily weighted toward water issues. Region 6 of the FWS administers two of the four LCCs containing greater sage-grouse habitat. In an effort to address larger-scale issues, Region 6 applied for funding from the national LCC Network for large-scale greater sage-grouse projects that transcend LCC boundaries. Although the region requested funding of $2.54 million spread over 3 years, it was successful in obtaining a grant of $500,000 from the LCC Network office. This money was used to solicit proposals for landscape-scale greater sage-grouse research and management.

The region issued a call for proposals for funding collaborative research and management support projects through WAFWA and Region 6 of the FWS, called the Inter-Landscape Conservation Cooperative (Inter-LCC) Greater Sage-Grouse Initiative.[6] Region 6 developed an agreement with WAFWA to collaboratively deliver two specific prongs of this initiative:

1. Funding support for priority greater sage-grouse research and management projects.
2. Supporting the identification and incorporation of existing data sets on greater sage-grouse and sagebrush ecosystems important to greater sage-grouse into the Landscape Conservation Management and Analysis Portal (LC MAP).

Several projects were funded through the initiative, including the following four projects:

- Range-wide sampling design for population size and trend estimation in greater sage-grouse;
- Sage-grouse hate trees: A range-wide solution for increasing bird benefits through accelerated conifer removal;
- Designing regional fuel breaks to protect large remnant tracts of sage-grouse habitat in Idaho, Oregon, Nevada, and Utah; and
- Forecast trends in sage-grouse populations by predicting future changes in habitat due to fire (Nevada) and climate change (Wyoming).

[6] See http://greatnorthernlcc.org/sites/default/files/lc_map-sage-grouse_rfp_final_v5.pdf.

Appendix B

Mississippi River Basin and Gulf Hypoxia: Collaborations Across Multiple LCCs

The Gulf hypoxia case study illustrates a response to addressing a conservation challenge that crosses Landscape Conservation Cooperative (LCC) boundaries, leading to collaboration among several LCCs, as in the sage-grouse study. It also illustrates an approach that is innovative, but challenging, in that it attempts to focus on an aspect of the problem not focused on by most of the current efforts and also that it adopts three fairly distinct conservation goals simultaneously (water quality, wildlife, and agriculture). A consortium of seven LCCs in the Mississippi River basin (Plains and Prairie Potholes, Upper Midwest and Great Lakes, Eastern Tallgrass Prairie and Big Rivers, Appalachian, Great Plains, Gulf Coast Prairie, and Gulf Coastal Plains and Ozarks) with funding from an LCC Network grant, developed the Mississippi River Basin/Gulf Hypoxia Initiative (MRB/GHI) (LCC Network, 2014).

The so-called Gulf dead zone, or region of hypoxia in the Gulf of Mexico off the mouth of the Mississippi River, has been a concern for some decades. Every summer, an area of between 2,000 and 8,000 square miles in which the level of dissolved oxygen falls below 2 parts per million forms off the Texas and Louisiana coasts (USGS, 2008). This low level of dissolved oxygen is known as hypoxia. This hypoxic zone leads to the death or outmigration of many aquatic organisms, including fish and shrimp, and has adverse ecological and economic consequences.

The formation of the zone is caused by nutrient-rich water coming out of the Mississippi River (NRC, 2012a). This nutrient-rich water promotes growth of plankton, which then dies, sinks out of the surface waters, and then decomposes, which reduces the oxygen. In addition, the freshwater from the river, being lighter than seawater, remains on the surface and causes stratification, which prevents new, oxygenated seawater from reaching the lower levels of the water column. The excess nutrients in the water derive mainly from agriculture in the Mississippi River basin.

Mitigating the Gulf hypoxia has been a goal for several programs at a variety of scales that are under way in the basin. The Mississippi River Gulf of Mexico Watershed Nutrient Task Force (MRGMWNTF), for example, was established in 1997 to "understand the causes and effects of eutrophication in the Gulf of Mexico; coordinate activities to reduce the size, severity, and duration; and ameliorate the effects of hypoxia." The Task Force includes the U.S. Environmental Protection Agency, the U.S. Department of Agriculture (USDA), the U.S. Department of the Interior, the U.S. Army Corps of Engineers, and the National Oceanic and Atmospheric Administration. In addition, it includes state agency representation from Arkansas, Illinois, Indiana, Iowa, Kentucky, Louisiana, Minnesota, Mississippi, Missouri, Ohio, Tennessee, and Wisconsin; and from the National Tribal Water Council. It has produced an Action Plan (MRGMWNTF, 2008)[1] and a reassessment (MRGMWNTF, 2013). The Action Plan "describes a national strategy to reduce, mitigate, and control hypoxia in the Northern Gulf of Mexico and improve water quality in the Mississippi River Basin." The reassessment concludes that although actions taken by the Task Force's members probably have helped to reduce the size of the hypoxic zone in recent years, environmental and economic conditions also have contributed. The National Research Council (NRC, 2012a) concludes that control efforts have not yet succeeded in reversing a long-term growth in the size of the hypoxic zone.

Another program is the Mississippi River Basin Initiative (MRBI), a "13-state initiative [that] builds on the cooperative work of NRCS and its conservation partners in the basin, and offers agricultural producers in priority watersheds the opportunity for voluntary technical and financial assistance."[2] The participating states are Arkansas, Illinois, Indiana, Iowa, Kentucky, Louisiana, Minnesota, Mississippi, Missouri, Ohio, South Dakota, Tennessee, and Wisconsin. This activity, also known as the Healthy

[1] See http://water.epa.gov/type/watersheds/named/msbasin/hypoxia101.cfm.

[2] See http://www.nrcs.usda.gov/wps/portal/nrcs/detail/oh/programs/landscape/?cid=stelprdb1119363.

Watersheds Initiative, was praised by the NRC in 2012 as having "attracted national-level attention and [holding] great potential. The USDA and its Natural Resources Conservation Service (NRCS) deserve recognition for its development and establishment."

The NRC has issued several reports on the problem of water quality in the Mississippi River Basin and hypoxia in the Gulf. The most recent (NRC, 2012a) report recommended that numeric nutrient criteria be established for the northern Gulf of Mexico's waters, saying that this would allow the U.S. Environmental Protection Agency and the Mississippi River states to work on developing allocations for nutrient loads throughout the basin. It also recommended establishing a basin-wide strategy for managing nutrients and water quality, adding

> [a]lthough there have been some federal and state efforts to coordinate nutrient management and related water quality programs across the Mississippi River basin, interagency efforts to date have not produced a rigorous, action-oriented plan for reducing nutrient loadings. There is no comprehensive river-basin wide program that includes, for example, interim water quality goals to be achieved over a specified time horizon, nutrient load allocations across the basin's tributary watersheds, a plan for more systematic data collection and analysis, or a framework of accountability to ensure achievement of goals and deadlines.

All the programs and effort described so far have focused entirely on water quality issues in the Gulf of Mexico. The effort reflected in the MRB/GHI aims to contribute through a slightly broader focus. As stated it is "intended to be complementary to related on-going efforts, like the Gulf of Mexico Hypoxia Task Force, Mississippi River Basin Initiative, and state nutrient reduction initiatives, but *with an added emphasis on the ecological and social values of wildlife habitat*" (emphasis in original). While still emphasizing water quality issues as central, this effort reflects a different, if not unique, perspective.

The area involved in the initiative is depicted in Figure B.1. The four fundamental objectives of the initiative are to increase fish and wildlife benefits, increase sustainable agricultural productivity, decrease Gulf hypoxia, and decrease implementation costs. The initiative focuses on four ecological systems, or focal habitats, and five agricultural production systems. The latter was done specifically in recognition that "practices will appeal differently to farmers producing a range of commodities." This is an example of recognition that conservation programs must be tailored to the people who will be affected for them to be effective.

The consortium generated a set of high-, medium-, and low-cost strategies (see Figure B.2). The group is using structured decision making to create an integrated framework to assess the strategies and to create a Landscape Conservation Design. This activity is fairly new, and so the committee's evaluation of the effort is focused on the approach rather than on the results to date.

The approach of this initiative has several positive features. As mentioned above, the recognition that ultimately conservation will be achieved by the work of people on the ground is a critical aspect of this initiative; this point also was made in a recent workshop on Mississippi River water quality and interstate collaboration, sponsored by the National Research Council (NRC, 2014a). Furthermore, the LCCs can contribute some research and monitoring components lacking in the MRBI. Another feature is the use of structured decision making to develop and prioritize strategies, taking costs into account. One is the recognition of ecological and wildlife values in addition to a more traditional focus on agricultural practices. Similarly, ecological restoration and conservation will need to be balanced with preservation of archeological sites and cultural values and preferences with regard to natural resource extraction. Obviously, the formation of a consortium of seven LCCs reflects an organic, adaptive flexibility in the LCC Network that seems likely to be effective and can assess the range of trade-offs.

For a summative evaluation in the future, it will be important to gather information and metrics that will enable one to answer questions such as the following: How well will the structured decision making translate into conservation gains and how well will different values be incorporated into the process? How well will the three disparate objectives (wildlife, water quality, and agriculture) be integrated and to what degree might they hinder progress or at least complicate evaluation of the initiative? Finally, how well will the MRB/GHI actually collaborate or at least coordinate with the several other Gulf hypoxia initiatives and activities currently under way?

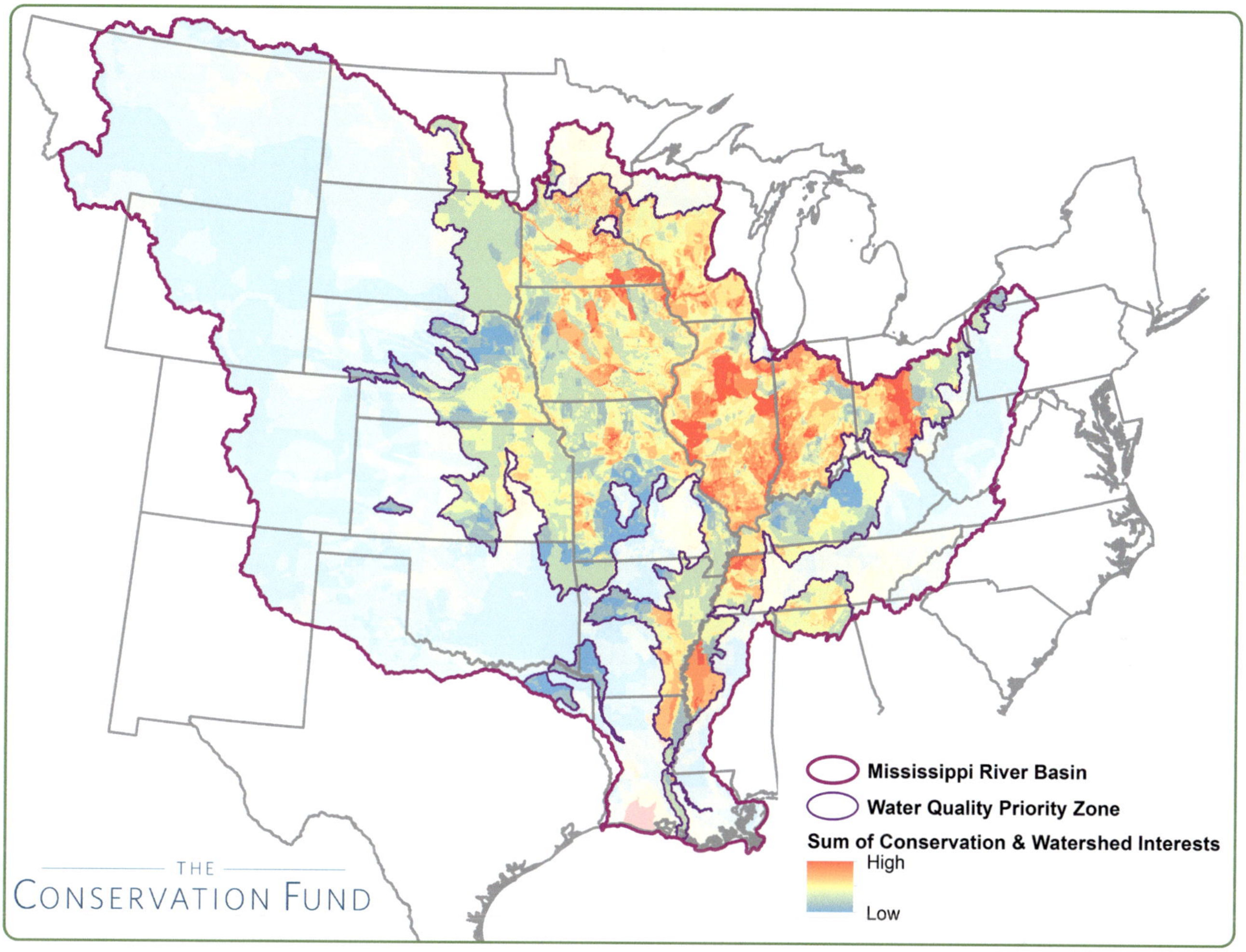

FIGURE B.1 The Mississippi River Basin/Gulf Hypoxia Initiative showing gradation of conservation interests at the intersection of wildlife, water quality, and agricultural production within the nutrient export priority zone for the multi-LCC Mississippi River Basin/Gulf Hypoxia Initiative (red = high; blue = low).
SOURCES: Eastern Tallgrass Prairie and Big Rivers LCC; the Conservation Fund.

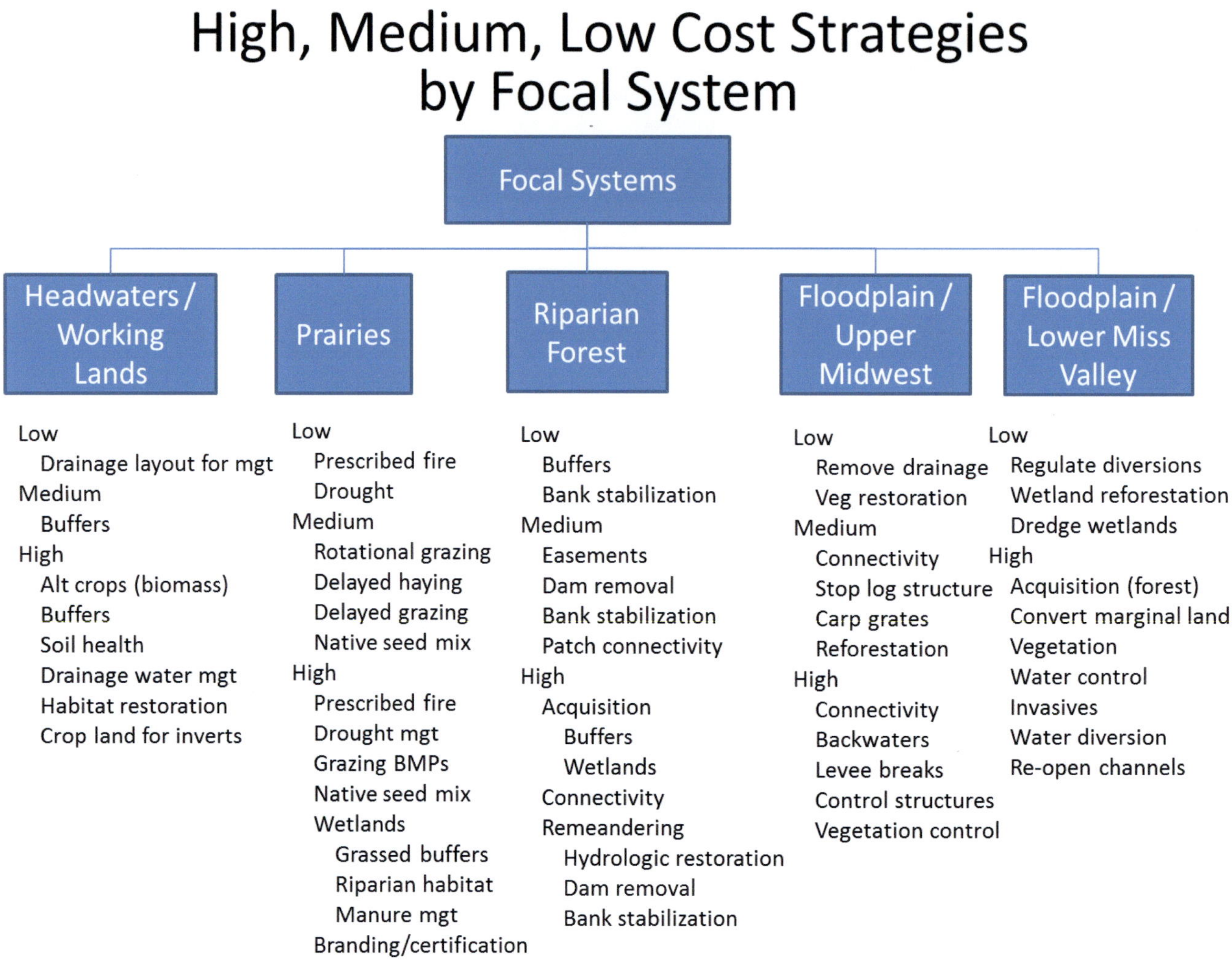

FIGURE B.2 High-, medium-, and low-cost strategies by focal system.
NOTE: alt = alternative; inverts = invertebrates; BMP = best management practice; mgt = management; veg = vegetation.

Appendix C

Guidance for Landscape Conservation Planning and Designs

LESSONS LEARNED IN DEVELOPMENT OF LANDSCAPE CONSERVATION PLANS

It will be helpful for Landscape Conservation Cooperative (LCC) planners and practitioners to be aware of some of the most important lessons learned and best practices in conservation planning (Groves and Game, 2015):

1. *Multiple objectives*. Early conservation plans were traditionally focused on a singular objective—conserving biodiversity (e.g., Groves et al., 2002). Today, most landscape conservation planning efforts will be trying to achieve multiple objectives, particularly those of the LCCs where so many different stakeholders and interests are engaged. Some of these objectives will be oriented toward biological features of the landscapes (targets), while other may be directed at cultural resources or ecosystem services. There are sophisticated planning tools for evaluating, comparing, and in some cases, conducting trade-off analyses between what may be competing objectives (e.g., Moffett and Sarkar 2006). There are now many examples of multiobjective planning within the conservation planning field. Marine spatial planning is one of the better known examples of planning for multiple objectives (Intergovernmental Oceanographic Commission, 2009).

2. *All forms of science*. Although traditional conservation plans focused primarily on the disciplines of ecology, wildlife biology, and conservation biology, many planners now appreciate that the disciplines of economics, social science, and political science will also make major contributions to landscape conservation plans. This situation is most easily recognized at the beginning of a planning effort when a team does a "situation analysis" that places the planning effort within the context in which it will occur and is increasingly referred to as the *socio-ecological system*—an acknowledgment that in most if not all landscapes the social, economic, and ecological systems are really one integrated system in which planning must be conducted (Ban et al., 2013).

3. *Integration of spatial and strategic planning*. Conservation planning was designed to answer two basic questions: Where should conservation take place on the ground (spatial planning) and how should it be achieved (strategic planning; Redford et al., 2003)? The two types of planning—spatial and strategic—are often conducted in separate processes that are poorly integrated. Spatial planning—or the location of areas important for biodiversity conservation—has dominated the field and we see evidence of that in the initial efforts in LCCs (i.e., the conservation blueprint-type maps). The most effective conservation plans are likely to be those that closely link the identification of places important for conservation with the strategies and actions necessary to achieve conservation (Game et al., 2013). Places for achieving conservation in and of themselves are not priorities; it is the actions that we need to take to conserve these areas that need to be prioritized because the actions take resources and we get various levels of return on investment for those actions and resources spent. The take-home message is that LCC landscape plans should spend just as much effort evaluating what types of strategies and actions to implement as identifying the areas for taking those actions.

4. *Evidence-based approach*. Thousands of conservation strategies and actions are being implemented every day—some succeed and some fail. At least two components of conservation planning can help in determining and implementing conservation strategies and actions that are more likely to succeed. The first of these is using a theory of change (see Figure C.1; CMP, 2013), in which the succeeding steps of a conservation strategy are diagrammatically represented so that the underlying logic of how the strategy is likely to succeed is examined through a series of steps in which each intermediate result from an action is portrayed in the results chain. The second is taking an evidence-based approach to selecting and implementing strategies and actions. In short, this means providing the evidence from experience or scientific literature that a particular conservation action is likely to succeed (Cook et al., 2010). For example, if a landscape

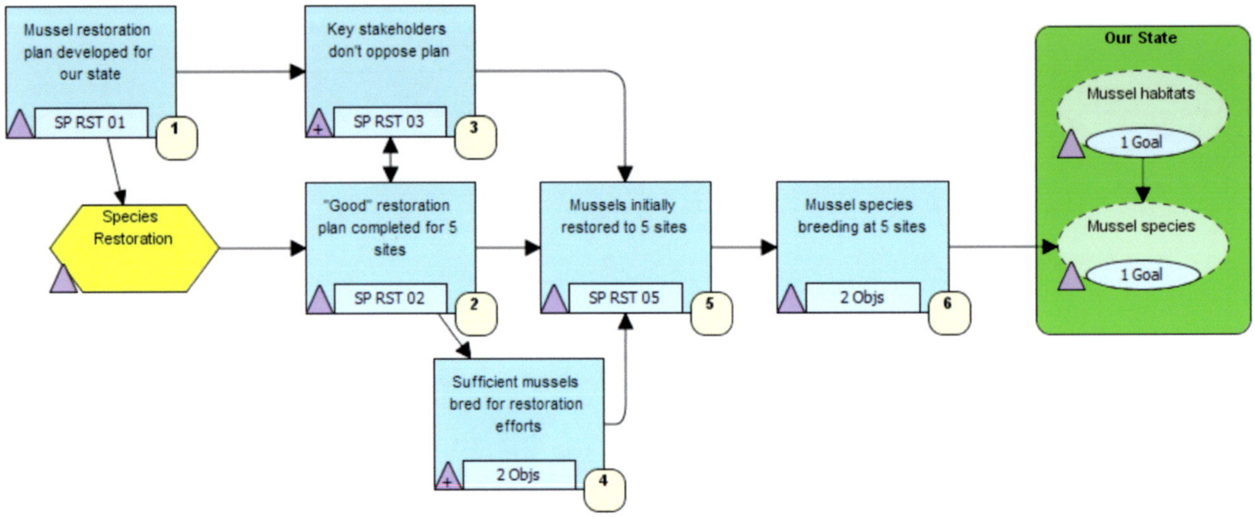

FIGURE C.1 Example of a theory of change or results chain diagram for implementing a specific conservation strategy and action. These diagrams help scientists and managers think through the implementation, monitoring, and evaluation of various conservation strategies and actions by illustrating the succeeding steps and underlying logic. They should be part of the overall Landscape Conservation Design and planning process. SOURCE: Used with permission from Conservation Measures Partnership (http://www.conservationmeasures.org).

plan suggests that a dam needs to be operated in a different fashion to provide more water downstream at particular times of year for endangered mussels, then there needs to be evidence evaluated and provided that such an action is likely to succeed.

5. *Using risk analysis*. Risks are simply uncertain events that could have a negative impact on the outcome of a conservation project. Certainly one of the underlying reasons for the creation of LCCs was the risk that climate change impacts could pose on the natural heritage of the United States. Yet, despite the fact that most scientists, planners, and practitioners know that there are various types of risk involved in conservation planning and implementation, few plans consider risk to any significant degree. The good news is that there are tried-and-true methods that can be integrated into conservation plans that will enable LCCs and others to better evaluate risks, and to ultimately implement strategies and actions in places that are more likely to succeed (Burgman, 2005).

6. *Planning-to-implementation gap*. A great deal has been written about the gap between developing a landscape conservation plan and implementing it (e.g., Knight et al., 2008). Because it remains unclear in these early years of the LCC Network whether the LCCs themselves are or can be an implementing body, any landscape conservation planning efforts need to pay particular attention to who the target audiences are for a conservation plan, and who will implement it, especially for strategies and actions that need to take place over multiple geopolitical jurisdictions. Here are a few tips to be on the lookout for to avoid some of pitfalls of poor implementation:

a. Many plans are simply too long and too detailed for stakeholders to understand. Shorter and succinct is better.
b. Too few plans give enough consideration to the financial and staff costs of implementing them. Money matters.
c. Some planning efforts lack engagement from those who will be charged with implementing them. It is an age-old adage but a person is more likely to be involved in solving the problem if he or she helps come up with the solution.
d. Some plans fail to articulate the conservation problem they are intended to address. Another way of thinking about this is that plans need specific objectives around which a set of prioritized conservation actions can be developed. "Prioritized" is the key word in this phrase; too many plans have long laundry lists of strategies and actions that lack a sense of priority and reality in the resources available to act on them. Plans that are more likely to succeed will spend considerable effort on developing a short list of high-priority strategies and actions and determining who will be responsible for them.

ILLUSTRATIVE COMPONENTS OF A REGIONAL CONSERVATION PLAN OR LANDSCAPE CONSERVATION DESIGN

Although all landscape conservation plans will take different forms depending on the desires of the stakeholders and the planning team, in one form or another the components outlined below should be part of the documentation process of any landscape planning effort. Different audiences will need varying levels of information about these components. How the plan is communicated will be a critical aspect of its success. Above all else, it is helpful to keep in mind why landscape conservation planning and design are so important: *it is to help ensure that conservation practitioners are making informed decisions on the strategies and actions that will help them achieve their ultimate conservation goals and objectives.*

- *The Executive Summary* is perhaps the most important section of the plan because many readers will not get beyond it; it's also a useful section for fundraising and outreach.
- *Planning context* includes the purpose of the plan, decisions to be made, decision makers, audience, constraints, or sideboards from previous planning efforts or law and policy.
- *Planning team and process* includes members, skill sets, organizations involved, team charter, management process, and roles.
- *Situation analysis* involves the economic, social, ecological, and political trends and opportunities within the socio-ecological system; it usually includes a conceptual model and assessment of threats to conservation features and may also include some analysis of enabling conditions for conservation and likely barriers to implementation.
- *Project scope* is the strategic, geographic, and temporal "boundaries" of the project.
- *Fundamental objectives and desired outcomes* include the ultimate outcomes in a conservation project that one hopes to achieve—the ends not the means—and those things that one cares most about.
- *Conservation features* are the elements of biodiversity, ecosystem processes, and social (human well-being) elements that are the focus of the planning efforts and, where appropriate, the quantitative targets (or goals) that have been set for these features.
- *The range of strategies* are the different strategies or major interventions that are under consideration for use in a conservation project or program and a rationale for how decisions will be made to focus on certain strategies and not others.
- *Strategy selection and theory of change* involve the strategies that a project or program has selected to implement and a rationale for how and why those strategies will be implemented.
- *Data and knowledge* are a summary of the types of data, knowledge (expert, local, traditional), and associated metadata that are used in the plan.
- *Risks* are those factors considered most likely to influence the successful implementation of strategies.
- *Monitoring program* is a plan for what actions will be taken during the project to measure progress and evaluate the effectiveness of strategies and actions.
- *Work planning* involves a detailed timeline of actions and tasks required to implement the plan, who is responsible, and proposed deadlines.
- *Budgeting and fundraising* involve detailed assessment of the staff and financial resources needed to implement the strategies and actions and a realistic fundraising plan to ensure that these resources are in place.
- *Communication* involves a summary of the different types of internal and external communications that will take place related to the project (e.g., websites, press releases, blogs, and field trips).
- *Operational or implementation plan* provides details on how the plan will be implemented.

COMMITTEE ANALYSIS

The illustrative components of a Landscape Conservation Design listed above will be critical components for the LCCs to include when developing their Landscape Conservation Designs. As discussed in great detail in Chapter 4 and to some extent in Chapter 6, developing metrics and approaches to account for on-the-ground conservation actions will be important, yet difficult for the LCCs. In contrast to the Joint Ventures, LCCs do not have the authority to deliver conservation actions. However, the LCCs can demonstrate how they contribute to on-the-ground conservation by developing—as part of the Landscape Conservation Design—a good theory of change, a monitoring program, and a clear work plan (see components, above).

Appendix D

Description of Other Federal Programs

This appendix describes other similar and related programs across the federal government, in less detail than in Chapter 5. Because of the number of existing programs, this review should not be considered comprehensive of all programs. Rather, the programs included here are programs the committee learned about during its analysis and are those that were commonly cited as potentially overlapping with the Landscape Conservation Cooperatives (LCCs) during the committee's information-gathering efforts.

Recognizing the importance of responding to a range of stressors including climate change in planning efforts at the sub-national scale (see Chapter 2), several federal agencies have created enterprises that serve a distinct set of stakeholders with information relevant to decision making in a changing world. While the missions of these enterprises emphasize climate to varying degrees, their processes for setting and achieving goals are quite different, as are their stakeholders, history, and institutional context. These enterprises share two important characteristics: an emphasis on regional (multistate) scale, and a convening function, by which the resource management agency is directed to work on the ground with regional partners.

The U.S. Department of the Interior (DOI) is a Cabinet-level agency comprising nine technical bureaus: U.S. Fish and Wildlife Service (FWS), Bureau of Indian Affairs, Bureau of Land Management (BLM), Bureau of Ocean Energy Management, Bureau of Reclamation, Bureau of Safety and Environmental Enforcement, National Park Service (NPS), Office of Surface Mining Reclamation and Enforcement, and the U.S. Geological Survey (USGS).[1] According to its mission statement, DOI "protects and manages the Nation's natural resources and cultural heritage; provides scientific and other information about those resources; and honors its trust responsibilities or special commitments to American Indians, Alaska Natives, and affiliated island communities."[2] LCCs and other similar programs, including those listed in Table 5.1, are examples of DOI's efforts to achieve this mission. Partnerships in particular have been a priority across DOI, and are credited with helping the Department achieve its mission and landscape-level conservation in particular.[3]

The FWS was created in 1940 during a reorganization of existing functions within DOI. The FWS today is a species and land management agency and a resource protection agency. Though the FWS directly manages 150 million acres through the National Wildlife Refuge System, the majority of freshwater fish and wildlife habitats occur on lands managed by other federal and state agencies or private landowners. Accordingly, the FWS has long recognized the need to partner with others to achieve conservation goals. Indeed, the agency's mission directly emphasizes partnerships ("Work with others to conserve, protect and enhance fish, wildlife and plants and their habitats for the continuing benefit of the American people."). The FWS has multiple ongoing programs organized across a variety of levels (national, regional, and state; see below).

NATIONAL PARK SERVICE SCALING UP

The NPS has particular expertise in cultural landscapes that can encompass a broad range of resources from designed landscapes, to large naturalistic parks, to living landscapes, and to landscapes that represent intangible values. These landscapes may be valued for the interaction of humans and their environment, for traditional cultural importance, for the traces of the past, and for the benefits that place may provide for the people of today. As documented in the recent publication *Scaling Up: Collaborative Approaches to Large Landscape Conservation*,[4] the NPS has a long history of working outside of traditional park boundaries with partnership parks, long-distance trails, and scenic river corridors. In

[1] See http://www.doi.gov/bureaus/index.cfm.

[2] See http://www.doi.gov/whoweare/Mission-Statement.cfm.

[3] See http://www.doi.gov//partnerships/index.cfm.

[4] See http://www.largelandscapenetwork.org/scaling_up.

response to the agency's strategic plan and DOI interest, the NPS has launched a special initiative around "Scaling Up," which includes a tool kit and a mapping program, NPScape.[5]

NPS's Scaling Up program is significantly different from the LCCs'. The work of Scaling Up is largely focused on preserving connectivity and access in the landscapes that contain national parks and protected areas. The work also more directly addresses cultural resources and cultural landscapes as part of the desired outcomes. Because of the focus on working to conserve resources on the ground through partnership networks, there is an emphasis on communications, forming a community of practice, and information sharing.

Although climate change is specifically addressed in the NPS's 2011 Call to Action (NPS, 2011)—for example, "the NPS is to be a leader in climate change adaptation in protected areas" (Action #21)—the link between research on climate change impacts and specific NPS landscape conservation strategies is not specifically addressed. For example, it is not referenced in the Chesapeake Bay Partnership, and climate change data were not referenced in the three available NPS examples.[6]

BUREAU OF LAND MANAGEMENT LANDSCAPE APPROACH

The Bureau of Land Management (BLM) administers 245 million acres of public land, more than any other federal agency, under a multiple-use mandate that includes extractive uses (e.g., oil and gas exploration, grazing, and timber harvests), recreation, and protection of natural, cultural, and historical resources. Most of the BLM-administered lands are located in the western states including Alaska.

The BLM Landscape Approach, which was designed to address the reality that public lands are influenced by challenges that transcend management boundaries, has been shaped by the experiences of scientists, land managers, and stakeholders who have sought to understand and address landscape-scale issues.[7] The approach is based around five components: (1) rapid ecoregional assessments (REA; synthesis of resource conditions and trends within a designated ecoregion), (2) ecoregional direction (identification of key management priorities for public lands within an ecoregion based on the REA and interactions with partners and stakeholders), (3) field implementation (adapting existing plans as needed and implementing management actions to achieve priorities), (4) monitoring for adaptive management (efforts are under way to increase the BLM's capability to monitor outcomes of management actions), and (5) science integration (integrating relevant science to inform management decisions, specifically citing the Climate Science Centers [CSCs] as providing information relevant to climate change).

The BLM defines landscapes as "large, connected geographical regions that have similar environmental characteristics."[8] The Landscape Approach encourages managers to consider the condition of natural resources and potential influences not only within their administrative unit but also within the context of the surrounding landscape with the goal of better understanding important ecological values and services as well as patterns of ecological change. The Landscape Approach is meant to provide a foundation for engagement with landowners and stakeholders and to inform resource management decisions made by field offices at the local level (field offices are an administrative unit with assigned jurisdictional boundaries where most management decisions are made).

The BLM indicates that the Landscape Approach supports the direction provided by DOI Secretarial Order No. 3289 (the same order that created the LCCs and CSCs) to more fully consider climate change in planning and decision making. While the concepts and experiences underlying this approach have developed over several years, the Landscape Approach program is relatively new (formalized in 2012).[9]

The BLM appears to view the LCCs as providing a mechanism to engage with important partners and stakeholders to implement their Landscape Approach. The BLM emphasizes the importance of partnerships to accomplish landscape-level management and references the national network of LCCs as DOI's effort to develop these partnerships.[10]

NATIONAL OCEANIC AND ATMOSPHERIC ADMINISTRATION REGIONAL COLLABORATION

The vision for the National Oceanic and Atmospheric Administration's (NOAA's) Regional Collaboration initiative is to provide "integrated services meeting the evolving demands of regional stakeholders." The collaboration represents the entire United States in eight regions and is staffed by NOAA employees and affiliates who promote coordination of the agency's regional capabilities and assets. Collaboration teams provide a systematic approach to internal coordination as well as external engagement. The Regional Collaboration Network promotes relationships, fosters communication around NOAA products and services, builds capacity for integrated products and services,

[5] NPScape, based on the NPS inventory monitoring program network, was developed as an assessment tool for managers of protected areas. It does not specifically address climate change although the data sets could be used for this purpose.

[6] See http://www.largelandscapenetwork.org/national-park-service/nps-case-studies.

[7] See http://www.blm.gov/wo/st/en/prog/more/Landscape_Approach.html#appr.

[8] See http://www.blm.gov/wo/st/en/prog/more/Landscape_Approach.html.

[9] See http://www.blm.gov/wo/st/en/info/regulations/Instruction_Memos_and_Bulletins/national_information/2012/IB_2012-058.html.

[10] See http://www.blm.gov/wo/st/en/prog/more/Landscape_Approach.html.

synthesizes regional trends, and recommends integrated solutions by engaging regional NOAA partners, stakeholders, and customers. Ultimately, regional team leaders bridge NOAA headquarters and its regional leadership, as well as provide a more regional context for integrative components of administration-wide messages. The Collaboration is guided by strategy based on environmental changes, performance results, prior-year performance, and administration priorities.[11]

U.S. DEPARTMENT OF AGRICULTURE

Regional Climate Hubs

The audience for the U.S. Department of Agriculture (USDA) Regional Climate Hubs (referred to herein as Climate Hubs) is almost exclusively private landowners (e.g., farmers, private forest owners, and ranchers).[12] The objective of the Climate Hubs program is primarily to "address risk management strategies on a regional basis, aiming to translate science and research, through extension and outreach, into actionable adaptation and mitigation practices for farmers, ranchers and forest landowners."[13] Moreover, the Climate Hubs program is nascent and small, with very little dedicated funding or staff. As such, the Climate Hubs program is unlike the LCC program in focus, audience, activities, and governance.

Collaborative Forest Landscape Restoration Program

The U.S. Forest Service (USFS) is an agency within the USDA that manages and protects 154 national forests and 20 grasslands that together total more than 192 million acres of public lands.[14] In 2009, through Title IV of the Omnibus Public Land Management Act of 2009, Congress established the USDA Collaborative Forest Landscape Restoration Program (CFLRP) to encourage the restoration of priority forest landscapes through collaboration and coordination with landowners. It also aims to encourage sustainability (ecological, economic, and social); leverage resources; reduce wildfire management costs; demonstrate the effectiveness of various restoration techniques; and encourage the use of restoration byproducts to offset the costs, benefit local communities, and improve forest health.[15]

The CFLRP selects and funds up to 10 restoration projects each year; however, the funding may only be used on National Forest System lands and may not cover planning costs.[16] To be eligible, projects must "have a landscape strategy, identify treatments for a ten-year period, be comprised primarily of National Forest System lands, reduce risk of uncharacteristic wildfire, reduce hazardous fuels, and encourage old growth."[17] They also must include a calculation of the cost savings.[18] Because the CFLRP is intended primarily to fund restoration projects and to do so primarily on National Forest System lands, it differs significantly from the LCCs in both mission and geographic scope.

Natural Resources Conservation Service Landscape Conservation Initiatives

The Natural Resources Conservation Service (NRCS) is an agency within the USDA whose mission it is to "provide resources to farmers and landowners to aid them with conservation."[19] Under the Food, Conservation, and Energy Act of 2008 ("2008 Farm Bill"),[20] the NRCS established landscape conservation initiatives. Through these initiatives, the NRCS strives to enhance locally driven and voluntary efforts in order to better address regional and national conservation issues. The initiatives extend beyond geopolitical boundaries and employ a science-based approach. Each landscape conservation initiative is focused on water quality and quantity, priority wildlife species, or ecosystems.[21] The NRCS landscape conservation initiatives are distinct from the LCCs in that their target audience is typically private landowners, and they are designed to support farmers, ranchers, and foresters in their efforts to simultaneously improve the environment and maintain a robust agricultural business. Also, while they are aimed at a regional scale, the NRCS initiatives are focused on particular resources and some overlap with one another geographically. Unlike the LCCs, the initiatives do not cover the entire United States.[22]

U.S. DEPARTMENT OF DEFENSE READINESS AND ENVIRONMENTAL PROTECTION INTEGRATION PROGRAM

The Readiness and Environmental Protection Integration program within the U.S. Department of Defense (DoD) aims to remove or avoid land-use conflicts near military installations. The purpose of this program is to ensure that military training, testing, and operations are not restricted or limited. An important aspect of this program is the use of buffer partnerships among the military services, conserva-

[11] See http://www.regions.noaa.gov/main.

[12] See http://climatehubs.oce.usda.gov/content/mission-and-vision.

[13] See http://climatehubs.oce.usda.gov/sites/default/files/hub-strategic-plan-draft-2015-03-23.pdf.

[14] See http://www.fs.fed.us/about-agency and http://www.fs.fed.us/about-agency/budget-performance.

[15] See http://www.fs.fed.us/restoration/CFLRP/overview.shtml.

[16] See http://www.fs.fed.us/restoration/CFLRP/overview.shtml.

[17] See http://www.fs.fed.us/restoration/CFLRP/overview.shtml.

[18] See http://www.fs.fed.us/restoration/documents/cflrp/July2010AdvisoryComMeeting/FinalJuly2010CFLRPMeetingSummary092310.pdf.

[19] See http://www.nrcs.usda.gov/wps/portal/nrcs/main/national/people.

[20] Pub. L. No. 110-234, 122 Stat. 923 (2008).

[21] See http://www.nrcs.usda.gov/wps/portal/nrcs/detailfull/nm/home/?cid=stelprdb1042113.

[22] See http://www.nrcs.usda.gov/wps/portal/nrcs/detailfull/nm/home/?cid=stelprdb1042113.

tion groups, and state and local governments, which acquire land and easements to preserve compatible uses and habitats near installations.[23] The program "also supports large landscape partnerships that advance cross-boundary solutions and link military readiness, conservation, and communities with federal and state partners through a common, collaborative framework."[24] With a focus on maintaining military operations, and geographic scope of the areas around military installations, the Readiness and Environmental Protection Integration program is unlike the LCCs.

OTHER U.S. FISH AND WILDLIFE SERVICE PROGRAMS

There are several potential areas of overlap between LCCs and other existing FWS programs whose purposes, broadly speaking, involve the development of applied science for conservation and/or convening partners. After reading about these programs, the committee determined that they were sufficiently different from the LCC program in scope, geographic scale, governance, topical focus, etc., that they do not overlap with the LCC program and did not merit further discussion:

- Conservation Planning Assistance Program,[25]
- Coastal Program,[26]
- National Wetlands Inventory,[27]
- Transportation Planning,[28]
- National Fish Hatchery System Science and Technology Program,[29]
- Wildlife Without Borders,[30]
- Aquatic Invasive Species Program,[31]
- Migratory Bird Program,[32] and
- Sport Fishing and Partnership Council.[33]

[23] See http://www.repi.mil/AboutREPI/FrequentlyAskedQuestions.aspx.
[24] See http://www.repi.mil/AboutREPI/FrequentlyAskedQuestions.aspx.

[25] See http://www.fws.gov/habitatconservation/cpa.html.
[26] See http://www.fws.gov/coastal.
[27] See http://www.fws.gov/wetlands.
[28] See http://www.fws.gov/transportationplanning.
[29] See http://www.fws.gov/fisheries/nfhs/science_tech.html.
[30] See http://www.fws.gov/international/wildlife-without-borders.
[31] See http://www.fws.gov/fisheries/ANS/ANS.html.
[32] See http://www.fws.gov/migratorybirds/SurveysMonitoringandResearch.html.
[33] See http://www.fws.gov/sfbpc.

Appendix E

Secretarial Order No. 3289

THE SECRETARY OF THE INTERIOR
Washington

ORDER NO. 3289, Amendment No. 1 ***(Amended material italicized)***

SIGNATURE DATE: February 22, 2010

Subject: Addressing the Impacts of Climate Change on America's Water, Land, and Other Natural and Cultural Resources

Sec. 1 **Purpose and Background.** Secretarial Order No. 3285, issued on March 11, 2009, made production and transmission of renewable energy on public lands a priority for the Department.

This Order establishes a Department-wide approach for applying scientific tools to increase understanding of climate change and to coordinate an effective response to its impacts on tribes and on the land, water, ocean, fish and wildlife, and cultural heritage resources that the Department manages. This Order replaces Secretarial Order No. 3226, Amendment No. 1, issued on January 16, 2009, and reinstates the provisions of Secretarial Order No. 3226, issued on January 19, 2001.

To fulfill our nation's vision for a clean energy economy, Interior is now managing America's public lands and oceans not just for balanced oil, natural gas, and coal development, but also—for the first time ever—to promote environmentally responsible renewable energy development. Sun, wind, biomass, and geothermal energy from our public and tribal lands is creating new jobs and will power millions of American homes and electric vehicles.

The Department is also taking the lead in protecting our country's water, land, fish and wildlife, and cultural heritage and tribal lands and resources from the dramatic effects of climate change that are already occurring—from the Arctic to the Everglades. The realities of climate change require us to change how we manage the land, water, fish and wildlife, and cultural heritage and tribal lands and resources we oversee. For example:

- New water management imperatives associated with climate change may require restoration of natural systems and construction of new infrastructure to reduce new flood risks or to capture early run-off.
- Strategies to address sea level rise may require acquisition of upland habitat and creation of wetlands and other natural filters and barriers to protect against sea level rise and storm surges. It may be necessary to relocate certain iconic and culturally historic structures.
- Shifting wildlife and habitat populations may require investments in new wildlife corridors.
- New invasions of exotic species and new wildland fire threats due to longer fire seasons and more severe droughts will require innovation and more effective ways of managing the Department's resources.

The Department of the Interior, with its 67,000 employees and scientific and resource management expertise, is responsible for helping protect the nation from the impacts of climate change. In particular the Department must:

- Adapt its water management strategies to address the possibility of shrinking water supplies and more frequent and extended droughts to continue to supply drinking water to more than 31 million people and irrigation water to 140,000 farmers.
- Wisely manage millions of acres of parks, refuges and other public lands, and prudently exercise its shared responsibility for managing the 1.7 billion acres of the U.S. outer continental shelf.
- Conserve and manage fish and wildlife resources, including over 800 native migratory bird species and nearly 2,000 federally listed threatened and endangered species.
- Protect cultural and archaeological resources and iconic structures that may be affected by climate change.

• Address the impacts of climate change on American Indians and Alaska Natives, for whom the Department holds trust responsibilities on behalf of the Federal government.

• Continue to provide state-of-the art science to better understand the impacts of climate change and to develop science-based adaptive management strategies for natural and cultural resource managers.

• Continue its work to quantify the amount of carbon stored in our forests, wetlands, and grasslands, identifying areas where carbon dioxide can be safely stored underground, and ways to reduce the Department's carbon footprint.

Sec. 2 **Authority.** This Order is issued under the authority of Section 2 of Reorganization Plan No. 3 of 1950 (64 Stat. 1262), as amended.

Sec. 3 **Coordinating the Department's Response to Climate Change Impacts on Our Resources.** ***The Climate Change Response Council within the Office of the Secretary is renamed the Energy and Climate Change Council (Council). The Council*** will execute a coordinated Department-wide strategy to address ***renewable energy efforts and*** to increase scientific understanding of and development of effective adaptive management tools ***to address*** the impacts of climate change on our natural and cultural resources. The ***Energy and Climate Change Council*** will be composed of the Secretary (Chair), Deputy Secretary (Vice-Chair), Counselor to the Secretary (Vice-Chair), Assistant Secretaries, Bureau Directors and the Solicitor. The Council will help coordinate activities within and among the Department's agencies and bureaus to develop and implement an integrated strategy for responding to ***renewable energy efforts and*** climate change impacts involving the resources managed by the Department. The Department's ***Energy and Climate Change Council*** will also coordinate its ***energy and*** climate change activities with all relevant Federal Departments and agencies including, but not limited to, the Council on Environmental Quality, the Office of Energy and Climate Change, the Office of Science and Technology Policy, the National Science and Technology Council, the Department of Agriculture, the Department of Commerce, the Department of Defense, and the Environmental Protection Agency.

The ***Energy and Climate Change Council*** will implement Department-specific ***energy activities as described in Secretarial Order # 3285 (Amendment No. 1), and implement*** climate change activities through the following mechanisms:

a. Climate Change Planning Requirements. Each bureau and office of the Department must consider and analyze potential climate change impacts when undertaking long-range planning exercises, setting priorities for scientific research and investigations, developing multiyear management plans, and making major decisions regarding potential use of resources under the Department's purview. These requirements were set forth in Secretary's Orders No. 3226 and 3285, and remain in effect. The organizational changes made by this Order will enable the bureaus and agencies to fulfill these planning requirements.

b. ***DOI Climate Science Centers.*** Management decisions made in response to climate change impacts must be informed by science and require that scientists work in tandem with those managers who are confronting climate change impacts and evaluating options to respond to such impacts. Pursuant to P.L. 110-161, the United States Geological Survey (USGS) has been developing regional science centers to provide climate change impact data and analysis geared to the needs of fish and wildlife managers as they develop adaptation strategies in response to climate change. These centers are currently known as "regional hubs" of the National Climate Change and Wildlife Science Center, and are being developed in close collaboration with Interior agencies and other federal, state, university, and non-governmental partners.

The Energy and Climate Change Council will work with USGS and other Department bureaus to rename these regional science centers as ***DOI Climate Science Centers (Centers)*** and broaden their mandate to encompass other climate-change-related impacts on Departmental resources.

These eight Centers will synthesize and integrate climate change impact data and develop tools that the Department's managers and partners can use when managing the Department's land, water, fish and wildlife, and cultural heritage resources.

c. Landscape Conservation Cooperatives. Given the broad impacts of climate change, management responses to such impacts must be coordinated on a landscape-level basis. For example, wildlife migration and related needs for new wildlife corridors, the spread of invasive species and wildfire risks, typically will extend beyond the borders of National Wildlife Refuges, BLM lands, or National Parks. Additionally, some bureau responsibilities (e.g., Fish and Wildlife Service migratory bird and threatened and endangered species responsibilities) extend nationally and globally. Because of the unprecedented scope of affected landscapes, Interior bureaus and agencies must work together, and with other federal, state, tribal and local governments, and private landowner partners, to develop landscape-level strategies for understanding and responding to climate change impacts. Interior bureaus and agencies, guided by the ***Energy and Climate Change Council,*** will work to stimulate the development of a network of collaborative "Landscape Conservation Cooperatives." These cooperatives, which already have been formed in some regions, will work interactively with the relevant ***DOI Climate Science Center(s)*** and help coordinate adaptation efforts in the region.

Sec. 4 **Additional Departmental Action to Mitigate Climate Change.** In accordance with Secretarial Order No. 3285, the Department has prioritized development ofrenewable energy on public lands and offshore waters to reduce our dependence on foreign oil and to reduce greenhouse gas pollution. This Order establishes two additional projects to mitigate climate change: the DOI Carbon Storage Project, and the DOI Carbon Footprint Project. Additional mitigation projects will be encouraged and supported by the ***Energy and Climate Change Council.***

a. The DOI Carbon Storage Project. This project is being implemented under P.L. 110-140, "The Energy Independence and Security Act of 2007," which gives the Department statutory responsibility to develop carbon sequestration methodologies for geological (i.e., underground) and biological (e.g., forests and rangelands) carbon storage. The USGS has the lead in administering the Carbon Storage Project, but will work closely with other bureaus and agencies in the Department and external partners to enhance carbon storage in geologic formations and in plants and soils in a manner consistent with the Department's responsibility to provide comprehensive, long-term stewardship of its resources. The DOI Carbon Storage Project is vital for successful domestic and global geological and biological carbon sequestration efforts.

b. The DOI Carbon Footprint Project. The project will develop a unified greenhouse gas emission reduction program, including setting a baseline and reduction goal for the Department's greenhouse gas emissions and energy use. The Assistant Secretary for Policy, Management and Budget will have the lead in administering the DOI Carbon Footprint Project, with the cooperation of all of the Department's agencies and bureaus.

Sec. 5 **American Indians and Alaska Natives.** Climate change may disproportionately affect tribes and their lands because they are heavily dependent on their natural resources for economic and cultural identity. As the Department has the primary trust responsibility for the Federal government for American Indians, Alaska Natives, and tribal lands and resources, the Department will ensure consistent and in-depth government-to-government consultation with tribes and Alaska Natives on the Department's climate change initiatives. Tribal values are critical to determining what is to be protected, why, and how to protect the interests of their communities. The Department will support the use of the best available science, including traditional ecological knowledge, in formulating policy pertaining to climate change. The Department will also support substantive participation by tribes in deliberations on climaterelated mechanisms, agreements, rules, and regulations.

Sec. 6 **Implementation.** The Deputy Secretary is responsible for ensuring implementation of all aspects of this Order. This responsibility may be delegated as appropriate. This Order does not alter or affect any existing duty or authority of individual bureaus.

Sec. 7 **Effective Date.** This Order is effective immediately and will remain in effect until its provisions are converted to the Departmental Manual or until it is amended, superseded, or revoked, whichever occurs first.

/s/ Ken Salazar
Secretary of the Interior

S0#3289Al 2/22/10

Appendix F

Landscape Conservation Cooperatives 2014 Network Strategic Plan

LANDSCAPE
CONSERVATION
COOPERATIVES

2014 | Network Strategic Plan

Landscape Conservation Cooperatives are public-private **partnerships** composed of states, tribes, federal agencies, non-governmental organizations, universities, international jurisdictions, and others **working** together to address landscape and seascape scale **conservation** issues.

ON THE COVER

River Otter in California. RICK KIMBLE/USFWS

Edwards Plateau region of Texas. TEXAS PARKS AND WILDLIFE DEPARTMENT

Tiger swallowtail butterfly on a thistle. JEFF BENNETT/NPS

Preamble

This strategic plan for the Landscape Conservation Cooperatives (LCC) Network builds on existing work within the Network and articulates a path for the next five years to achieving the LCC Network's vision and mission.

The purpose of the LCC Network is to harness the capacities and abilities of all partners in support of common conservation outcomes and to serve as a strategic forum for collegial collaboration, coordination, and integration. The Network is an extension of existing partnership efforts, which seeks to enhance conservation outcomes across the geographical extent of the Network.

Individually, each LCC is a collaborative, self-directed conservation partnership that connects partners and pre-existing local partnerships to a landscape vision. LCCs undertake work specific to the needs of their geographies and collaborators.

The LCC Network Strategic Plan identifies goals, objectives, and example tactics that support the Network's vision and mission. The goals identify common aims of individual LCCs and provide a way for them to align across the Network's geography. Each of the four goals—addressing **conservation strategy**, **collaborative conservation**, **science**, and **communications**—has a set of objectives.

The objectives describe LCC efforts that will be prioritized and implemented differently according to each LCC's unique characteristics. Example tactics are listed for each objective purely for illustration. They do not prescribe what the self-directed LCCs "should" be doing. Rather, they are examples provided by the LCC community that show the diversity of conservation approaches; there is no expectation or direction that all LCCs will use all tactics.

This strategic plan

- » Provides a living, iterative reference to guide the LCC Network[1]
- » Builds from existing strategies within the LCC Network—every LCC should see their work somewhere in the document
- » Does not prioritize any goal, objective, or example tactic as more important than another
- » Will be updated in the next two years

The strategic plan does not

- » Prescribe the actions individual LCCs must take to achieve the vision and mission of the LCC Network
- » Encompass all things that all LCCs are working on
- » Make value statements on priorities for individual LCCs

This strategy contains common and shared elements to help individual LCCs work collaboratively as a conservation network while communicating how the broader LCC Network collectively functions.

The information and ideas in this strategic plan were gleaned from existing LCC strategic, science, operational and other plans. This strategic plan serves as a companion document to these plans. The original draft of this document resulted from a weeklong workshop in late July 2014 of over 50 representatives from within the LCC Network, including members of the LCC Council, Steering Committees, LCC Coordinators, LCC Science Coordinators, national partner organizations, and other Network leaders. The original draft was then improved by substantive comments from more than three dozen organizations within the LCC Network.

LCCs are fundamentally partnership endeavors. Although LCCs appear to be a new construct, they actually are based on existing models. The LCC Network appreciates and celebrates the long-term partnerships that have preceded its establishment and have helped steer the Network towards this collaboration model. These long-term partnerships are truly the foundation and models for success. Conservation partners, whether or not engaged with LCCs, produce valuable conservation science and delivery that intertwines with LCC goals. This strategic plan supports and does not abrogate or diminish the authorities and responsibilities of partners. Through individual partner responsibilities, authorities and accomplishments, the Network will achieve success. It is anticipated that partners will engage in those goals and objectives that are important to them and that align with their unique mission.

1 Some sections include specific key or technical terms. Definitions for some of these have been provided. Readers of this strategic plan should think about these terms in the context of their own geography, community, culture or organization. The meaning of these terms can change over time through the efforts of a collaborating and evolving community.

Vision: Landscapes capable of sustaining natural and cultural resources for current and future generations

LCC Network

Sunrise in the desert.
ANDREW LOESCHER/USFWS

MISSION

A network of cooperatives depends on LCCs to:

» Develop and provide integrated science-based information about the implications of climate change and other stressors for the sustainability of natural and cultural resources;

» Develop shared, landscape-level, conservation objectives and inform conservation strategies that are based on a shared scientific understanding about the landscape, including the implications of current and future environmental stressors;

» Facilitate the exchange of applied science in the implementation of conservation strategies and products developed by the Cooperative or their partners;

» Monitor and evaluate the effectiveness of LCC conservation strategies in meeting shared objectives;

» Develop appropriate linkages that connect LCCs to ensure an effective network.

Introduction

The purpose of this Landscape Conservation Cooperative (LCC) Network Strategic Plan is to create a strategic framework with shared goals for how the LCCs will aspire to achieve the LCC Network vision through collective impact.

The success of the Network depends both upon the success of individual LCC self-directed partnerships in addressing the conservation needs most important within their geographies, as well as the LCCs' collective ability to address conservation goals at even broader geographic scales.

Recognizing that conservation objectives (shared or unique) are achieved at the scale of individual LCCs, monitoring the Network's effectiveness should reflect the successes of each LCC and Network accomplishments. As an assemblage of conservation practitioners across the continent, the Pacific Islands and the Caribbean, LCCs have an obligation to future generations who will rely on natural and cultural resources for their livelihood, quality of life, and cultural connection. The network of LCCs works collectively to conserve and maintain landscapes and seascapes[2] capable of sustaining natural and cultural resources for current and future generations.

The LCC Network's aspiration to develop an ecologically connected network of landscapes and seascapes can be achieved through the identification and pursuit of shared, broad-scale conservation goals that span political, jurisdictional, and ecological boundaries, along with a shared understanding of the problems the Network is seeking to address. The broad geographic scope of the Network is necessary to facilitate and support unprecedented integration efforts and mechanisms that address large-scale stressors such as climate change, urbanization, pollution, energy development, resource extraction, and water stress. Working collectively, LCCs across the entire Network can face complex problems that no single organization or entity can solve alone.

This LCC Network Strategic Plan encourages national and international leaders to support the Network through information, policy, funding, and conservation action. It describes some of the shared goals and objectives that could work from multi-LCC to international scales, yet it recognizes that the success of the LCC Network depends upon the conservation successes of individual LCCs.

2 The term "landscapes and seascapes," or sometimes just "landscapes," hereafter refers to terrestrial, aquatic and marine environments along with a supportive, healthy natural environment with clean air and water.

Goals & Objectives

An actionable strategy to adapt conservation to a changing world.

Children planting at Desert National Wildlife Refuge in Nevada. USFWS

The LCC Network Strategic Plan is organized into four strategic goal areas:

1. **Conservation strategy**
2. **Collaborative conservation**
3. **Science**
4. **Communications**

Each strategic area has a goal, a set of objectives, and example tactics identified under those objectives (a longer list of example tactics is included in the Appendix). The example tactics serve to illustrate what can be done to achieve the objectives, at multiple levels and degrees of specificity, which can be refined during implementation. They are not inclusive of all future actions potentially needed to meet these objectives.

1 Conservation Strategy

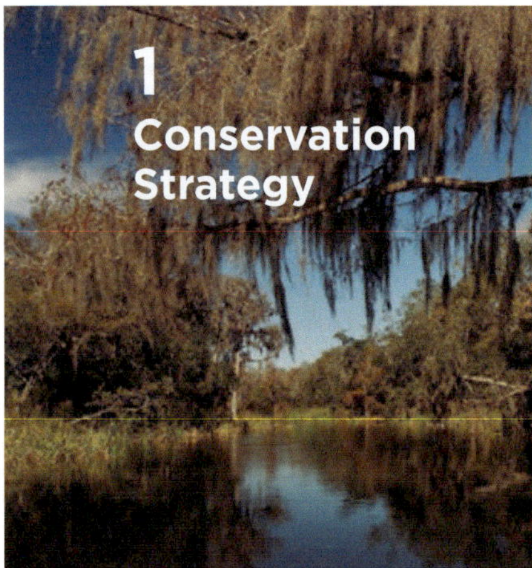

GOAL 1

An ecologically connected network of landscapes and seascapes adaptable to global change—such as climate change—with the ability to sustain ecological integrity and health to meet the needs of society at multiple scales.

Left: Wacissa River in Florida. FLORIDA FISH & WILDLIFE CONSERVATION COMMISSION **Right: Wood thrush.** STEVE MASLOWSKI/USFWS

Objective 1

Identify shared conservation objectives, challenges, and opportunities to inform landscape conservation at continental, LCC, island, and regional scales.

EXAMPLE TACTICS

» Establish conservation objectives at the LCC level and other applicable scales.

» Roll-up LCC objectives to identify Network-scale objectives.

Objective 2

Develop then deliver (through partners) regional landscape conservation goals and designs[3] that support resiliency and adaptation to both global change and regional landscape challenges, while ensuring the inclusion of all partners and stakeholders necessary for successful conservation.

EXAMPLE TACTICS

» Work with all necessary partners and indigenous peoples to select flagship regions and identify priority areas for conservation and restoration.

» Complete, expand, and support the delivery of landscape conservation designs by providing tools and guidance to assist in their implementation.

3 Landscape Conservation Design is an iterative, collaborative, and holistic process that provides information, analytical tools, spatially explicit data and best management practices to develop shared conservation strategies and to achieve jointly held conservation goals among partners.

Healthy lands support diverse wildlife populations for wildlife watchers, anglers and hunters. USFWS

» When developing conservation designs, acknowledge the extensive knowledge and practices indigenous peoples have developed over generations about large landscape processes.

Objective 3

Integrate regional or other scale-specific conservation designs to align and focus conservation action at the Network scale, within available authorities.

EXAMPLE TACTICS

» Facilitate the design of an ecologically connected network of large geographic regions that support priority natural and cultural resources.

Objective 4

Identify and obtain the resources required at the LCC and Network scales to inform, develop, and support implementation of the conservation designs and other conservation actions.

EXAMPLE TACTICS

» The LCC Council and Network partners will identify existing and new resources, then advocate for and pursue the resources required to accomplish this goal at the LCC Network level.

Objective 5

Monitor the effectiveness of conservation design(s) and design application in terms of achieving stated outcomes, then revise as appropriate.

EXAMPLE TACTICS

» Identify and promote good, tested practices for design development and supporting implementation.

2 Collaborative Conservation

GOAL 2

Facilitated alignment of partnership efforts within and amongst LCCs, including planning efforts and resources, that improves conservation outcomes across LCCs and the Network.

Left: Collaborators from Caribbean LCC. OLIVER BENCOSME/SEA GRANT PUERTO RICO **Right: Hurricane Irene damage on Pea Island.** TOM MAKENZIE/USFWS

Objective 1

Create a high-functioning organizational culture for LCCs and the Network.

EXAMPLE TACTICS

» Identify institutional barriers and stovepipes that inhibit cross-agency collaboration and partnerships and seek to reduce, break-down, or overcome them.

» Continue to pursue opportunities to expand partnerships throughout the Network, including the addition of new partners in LCCs—particularly nontraditional partners that increase the breadth, diversity, and effectiveness of the conservation community.

Objective 2

Identify and explore opportunities for collaborative actions within the LCC Network.

EXAMPLE TACTICS

» Identify conservation and mitigation opportunities that span multiple LCCs as well as those that extend to other regional collaboration networks.

Spring Beauty at Sunflower Flats, Elko District BLM Nevada. SHANELL OWEN

Objective 3

Demonstrate, monitor, and evaluate the value and effectiveness of the LCC Network.

EXAMPLE TACTICS

» Develop common definitions and performance metrics for key qualitative and quantitative outcomes that highlight, show, and demonstrate value.

Objective 4

Identify new and existing training and networking opportunities for the LCC Network.

EXAMPLE TACTICS

» Create a "playbook" containing best management practices for LCC Coordinators and Steering Committee Chairs.

Objective 5

Leverage conservation planning to be opportunistic in taking advantage of current and new funding sources for conservation.

EXAMPLE TACTICS

» Prepare to respond to funding and partnership opportunities that arise as a result of urgent conservation needs (e.g. natural disasters, species invasions, disease) that are likely to occur based on future scenario planning.

Objective 6

Create a Network-level system for prioritizing operational needs at Network and regional levels, as appropriate.

3 Science

GOAL 3

Natural and cultural resources are conserved at large landscape and seascape scales, guided by the collaborative application of science, experience, and cultural or traditional ecological knowledge and the generation of new conservation knowledge.

Left: Aurora borealis at Sherburne National Wildlife Refuge. BRYAN WORTH. Right: Magnificent Frigatebirds. USFWS

Objective 1

Identify shared science, information, and resource needs at the Network-scale.

EXAMPLE TACTICS

» Complete, disseminate, and implement the LCC Network Science Plan.

» Identify and develop critical spatial, biological, and cultural data and evaluation tools across the Network.

» Support assessment of climate change impacts and adaptation planning for cultural and/or subsistence resources that are traditionally gathered, hunted, or culturally significant.

Objective 2

Promote collaborative production of science and research—including human dimensions—as well as the use of experience and indigenous and traditional ecological knowledge among LCCs, Climate Science Centers (CSCs), and other interested parties; use these to inform resource management decisions, educate local communities, and address shared needs.

Child with fish at the Upper Mississippi River National Wildlife Refuge.
BOB DRIESLEIN/ USFWS

Traditional ecological knowledge[4] refers to the knowledge, innovations, and practices of indigenous and local communities around the world.

EXAMPLE TACTICS

» Engage resource decision makers, managers, cultural practitioners, and indigenous peoples in the appropriate framing of resource-based decision problems and the formulation of clear management objectives that focus and guide subsequent science activities.

Objective 3

» Demonstrate and evaluate the value and improve the effectiveness of LCC science.

EXAMPLE TACTICS

» Support efficiency among LCC and other appropriate broad-scale monitoring programs in generating status and trend information on priority resources and landscapes by facilitating sharing, cooperative synthesis, communication, and evaluation of data.

4 Traditional ecological knowledge is developed from experience gained over the centuries and adapted to the local culture and environment. The word "knowledge" is meant to be plural, to acknowledge the many types of knowledge it includes.

4 Communications

GOAL 4

Advance the knowledge of, support for, and engagement in landscape-scale conservation across the LCC Network.

Left: A school of manini at Kingman Reef National Wildlife Refuge. KYDD POLLOCK
Right: Atlantic puffins at Maine Coastal Islands National Wildlife Refuge. USFWS

Objective 1

Communicate the existence and application of LCC Network science, products and tools to partners and stakeholders in a form that is understandable, publicly accessible, engaging, and relates to what matters to end users and society.

EXAMPLE TACTICS

» Use the LCC Network website and other tools as a platform for sharing key LCC services and benefits such as news, products, tools, training, science, data, documents, and open source software for conservation use or research, where appropriate.

Objective 2

Increase two-way communication with, outreach to, and engagement of key partners across the LCC Network as well as new partners to expand the LCC Network and increase conservation impact and achievements.

EXAMPLE TACTICS

» Identify new, strategic target audiences with interests that may intersect with conservation, such as young leaders, environmental justice groups, development communities, resource extraction industries, or planners from the built environment who could use science to make decisions that support both conservation and their own interests.

Paddling the canals of Alligator River National Wildlife Refuge. STEVE HILLEBRAND, USFWS

Objective 3

Develop and implement a communications and outreach plan that identifies and uses media to clearly convey to appropriate target audiences the value and tangible successes of the LCC Network at various scales.

EXAMPLE TACTICS

» Communicate LCC successes that show on-the-ground impact or how they have influenced management decisions through a variety of means. These successes also need to convey the complexities and challenges of achieving landscape- and seascape-scale conservation results.

» Open and increase the availability of LCC products and information to diverse audiences.

Objective 4

Build communications capacity and capabilities within the LCC Network to effectively communicate the purposes and successes of the LCC Network.

EXAMPLE TACTICS

» Foster regional communications communities of practice that build upon the strengths and expertise of LCCs and their individual partners.

Objective 5

Share lessons learned across the LCC Network.

Putting It Into Action

As a living, iterative document, this strategic plan will guide LCC actions to achieve the Network vision and mission. These actions, however, will vary depending on geographic differences and other variations among individual LCCs.

A full list of the example tactics generated through this strategic planning process is included in the Appendix, but it was never considered to be a complete list of tactics. The LCC Network governance infrastructure, including the LCC Council, LCT, LSCT, steering committees of individual LCCs, and other entities within the Network, can select which issues best intersect and meet their own needs as well as the needs of the Network. Partners are invited to collaborate and identify the best approaches for action, measurable outcomes, and needed resources to successfully implement strategies outlined in this plan.

Implementing this strategic plan will require the development of an implementation framework. The framework will identify processes to advance the goals and objectives in this strategy and will establish a schedule for monitoring and evaluating program performance.

Working groups or teams may be established to formulate next steps and help monitor and communicate how the Network collectively is addressing specific goals and objectives. The LCC Network plans to assign an LCC Strategic Plan Implementation Coordinator to facilitate broader communication across the Network. The LCC

> To borrow Aldo Leopold's thoughts on the land ethic, nothing as important as the future for the LCC Network is ever 'written'—it evolves in the minds of a thinking, collaborative community.

Left: Red foxes on Edwin B. Forsythe National Wildlife Refuge. DON FREIDAY **Middle: Beach on the coast of Oregon.** JOHN MANKOWSKI/NPLCC **Right: Flint Hills Legacy Conservation Area, Kansas.** RICK HANSEN/USFWS

Network, with coordination from the LCC Network staff, will gather information about lessons learned and facilitate an ongoing dialogue with the intent to update the LCC Network Strategic Plan within the next two years.

The LCC Network as an entity will continue to listen, evolve, and support a diverse array of partners all working together to fulfill its vision for landscapes capable of sustaining natural and cultural resources for current and future generations.

Appendix

Each goal area has a set of objectives and some example tactics, which are intended to help illustrate the kinds of actions that can be taken to help achieve the Network's common goals.

This appendix includes a list of all the example tactics generated through the strategic planning process but this list is not meant to limit the tactics available to the Network. The actual tactics deployed will be determined by the various entities within the LCC Network (see "Definitions" section of this Appendix) based on which actions they support and the implementation framework that will be developed for this strategic plan.

GOAL 1: CONSERVATION STRATEGY

Objective 1: Shared Objectives

» Establish conservation objectives at the LCC level and other applicable scales

» Roll-up LCC objectives to identify Network-scale objectives

Objective 2: Conservation Designs

» Work with all necessary partners and indigenous peoples to select flagship regions and identify priority areas for conservation and restoration with the aim of facilitating climate resilient lands and waters; build, maintain, or restore resilience in vulnerable regions; develop or increase carbon storage capacity (where appropriate); and address management issues within given authorities.

» Produce first generation climate change resilient landscape designs.

» Complete, expand, and support the delivery of landscape conservation designs with goals for priority areas and provide tools and guidance to assist in their implementation.

» When developing conservation designs, acknowledge the extensive knowledge and practices indigenous peoples have related to large landscape processes.

» Identify priority areas where opportunities exist to improve resilience or adaptation strategies for priority resources, ecosystem services, and communities.

» Support the development of foundational data sets at the LCC scale that could also be rolled up to the multi-LCC, continental, and global scales.

» Support development of analytical tools to help understand the effects of global change on natural and human systems at the LCC, multi-LCC, continental, and global scales.

» Identify and engage Network-wide common partnerships that should be engaged in support of this objective (e.g., National Association of Counties).

Ecological resilience is the capacity of a system to resist and recover from natural or human-cause disturbances. Resilient systems can maintain their essential structure in the face of floods, fires, pest outbreaks, pollution, and other stressors.

Objective 3: Integration

» Facilitate the design of an ecologically connected network of large geographic regions that support priority natural and cultural resources.

» Identify, prioritize, and support implementation of cross-LCC actions where coordinated action across several LCCs could have a multiplying effect.

» Support the implementation of national plans to achieve landscape conservation and climate adaptation such as the National Fish, Wildlife & Plants Climate Adaptation Strategy, particularly as it relates to other goals and objectives of this strategic plan.

» Develop strategies to address cumulative, existing (e.g., non-climate) stressors broadly and major global change stressors at the relevant scale.

» Identify additional strategies needed to adapt to global change that incorporate human and societal values.

» Recognizing the proprietary status of traditional ecological knowledge, identify Tribal and First Nations' information needs related to conservation and management of natural and cultural resources potentially affected by global change, including climate change.

» Collect, evaluate, analyze, then provide the best tools and information that managers need to assist them in supporting design implementation and evaluation.

Objective 4: Needed Resources

» Identify the core resources—including for staff, science, partners, and community assets—needed for each LCC and the shared resources and approaches that would increase appropriately consistent, but not identical, approaches and integration between LCCs to accomplish this strategic goal Network-wide.

» The LCC Council and Network partners will identify existing and new resources, then advocate for and pursue the resources required to accomplish this goal at the LCC Network level.

» Develop a set of transparent performance metrics that enable the Network to monitor and evaluate progress.

Objective 5: Monitoring Effectiveness

» Identify and promote good, tested practices for design development and supporting implementation.

» Design a process or framework for evaluation of conservation designs' success using shared methodologies, standards, and other approaches across the Network, collectively giving a sense of Network-level effectiveness.

» Develop the needed foundation that status and trend monitoring provide, in addition to existing inventory monitoring and data sets.

GOAL 2: COLLABORATIVE CONSERVATION

Objective 1: Organizational Culture

» Identify institutional barriers and stove-pipes that inhibit cross-agency collaboration and partnerships and seek to reduce, breakdown, or overcome them.

» Create a forum for regular communication across LCCs and throughout the Network.

» Revise, as needed, the Network's organizational structure along with defined relationships, roles, and functions within the Network.

» Continue to pursue opportunities to expand partnerships throughout the Network, including the addition of new partners in LCCs—particularly nontraditional partners that increase the breadth, diversity, and effectiveness of the conservation community.

» Increase ownership, participation, and engagement of partners in the work of the LCCs through developing common approaches, sharing tools, assigning leads for important products, pooling resources, and other such approaches.

» Continue to provide and develop additional funding mechanisms to facilitate key partner engagement.

Objective 2: Collaboration Opportunities

» Identify commonalties (e.g., needs and tools) and differences that require solutions.

» Encourage the establishment of partner forum events that connect conservation partners to local on-the-ground action and that connect national or regional organizations to actions and initiatives at larger scales.

» Identify inter-LCC conservation goals, challenges (e.g., data gaps, policy, technical), and opportunities.

» Identify successful inter-LCC collaboration efforts, and then celebrate, communicate, and share them.

» Identify conservation and mitigation opportunities that span multiple LCCs as well as those that extend to other regional collaboration networks.

» Identify opportunities to collaborate on mitigation activities and methods.

» Identify, celebrate, and share good practices and success stories.

Objective 3: Network Effectiveness

» Develop common definitions and performance metrics for key (qualitative and quantitative) outcomes that highlight, show, and demonstrate value.

» Measure collective impact (e.g., resiliency, footprint, connectivity) of those outcomes and others when needed.

» Identify the most appropriate metrics with which to measure partner inputs into Network operations and projects (e.g., financial, personnel, technical assistance, in-kind) and capture outcomes achieved indirectly by the LCCs, when feasible.

» Develop accountability tools.

» Explore ways for partner organizations to maximize the benefits of performance reporting within and amongst their organizations, including at the Network level.

Objective 4: Training and Networking

» Provide orientation training for new LCC staff and steering committee members.

» Create a "playbook" containing best management practices for LCC Coordinators and Steering Committee Chairs.

» Develop and distribute a "communications resource" guide for Steering Committee and LCC Council members to help effectively communicate the LCC Network vision and actions across all levels of their organizations.

» Identify or produce training for LCC Coordinators related to developing management and coordination skills.

Objective 5: Funding Opportunities

» Develop proactive systems, protocols, and strategies for identifying and prioritizing high-priority conservation needs (both natural and cultural) in the event resources become available in relation to an established baseline. Determine how to fund rapid response opportunities to urgent conservation needs (e.g., natural disasters, species invasions, disease) and ensure the necessary capacity.

Objective 6: Prioritizing Investments

» Inventory conservation needs, planning efforts, and resources.

» Identify which efforts would maximize the network-wide conservation benefit of investments.

» Balance resource investments with LCC needs and strategies for improving performance.

GOAL 3: SCIENCE

Objective 1: Identifying Needs

» Complete, disseminate, and implement the LCC Network Science Plan.

» Inform and articulate network-wide or regional conservation targets (and their associated goals and objectives) reflective of the vision to achieve an ecologically connected landscape.

» Identify and develop critical spatial, biological, and cultural data and evaluation tools across the Network.

» Identify knowledge gaps and define research priorities via a collaborative process with federal, state, tribal, private conservation organizations, academic resource managers and research scientists.

» Leverage resources to conduct focused research to fill critical knowledge gaps in conservation science.

» Support assessment of impacts and adaptation planning for cultural and/or subsistence resources that are traditionally gathered, hunted, or culturally significant

» Share guidelines developed by indigenous working groups for integrating scientific and traditional ecological knowledge within and between LCCs.

» Partners within LCCs educate other partners at the table about each agency's needs, interests, and programs.

Objective 2: Co-production

» Engage resource decision makers, managers, cultural practitioners, and indigenous peoples in the appropriate framing of resource-based decision problems and the formulation of clear management objectives that focus and guide subsequent science activities.

» Engage these same groups and other end users in the identification, development, production, and use of scientific tools through technical assistance, outreach, training, and education.

» Promote funding notifications to Tribes and First Nations for the documentation and exchange of traditional ecological knowledge with scientists to generate solutions through co-learning and co-production of knowledge.

» Promote continual learning to improve conservation science by sharing good, tested practices and standardized approaches.

» Provide Network-wide context for connecting the science efforts of individual LCCs to ensure they reflect the needs of LCC priority resources across the landscapes and seascapes.

» Create forums for scientists to work with indigenous peoples to co-produce knowledge that can address landscape-scale issues in accordance with free, prior, and informed consent.[5]

» Improve technical approaches for management and integration of adjacent conservation designs to foster a seamless, resilient, and interconnected ecological network of lands and waters.

» Translate shared conservation goals and science through spatial products and other tools to guide action at the landscape scale in those geographies where these products are value-added.

Objective 3: Evaluating Science

» Develop metrics and methods to evaluate the use of LCC science and monitoring information in shaping resource planning, management decisions, and community adaptation.

» Develop processes and tools that allow the LCC Network to measure progress toward achieving and retaining resilient and functional landscapes and seascapes.

» Support efficiency among LCCs and other appropriate broad-scale monitoring programs in generating status and trend information on priority resources and landscapes by facilitating sharing, cooperative synthesizing, communications, and evaluation.

» Improve the efficiency of conservation design and delivery process.

» Demonstrate the iterative nature of the questions and issues that resource managers face and the resulting determinations of resource priorities for shared needs.

» Assign project support to Tribes and First Nations to demonstrate examples of how, when, and where traditional ecological knowledge can be used to better inform management decisions.

5 Free, prior, and informed consent is a principle that means an indigenous group has the right to share or withhold information or traditional ecological knowledge in accordance with their beliefs, customs, rules, and traditions.

GOAL 4: COMMUNICATIONS

Objective 1: Communicating Science

» Encourage communications guidance, policy, training, and support to principle investigators for science delivery regarding outreach strategies and applications of their research and results to end users (e.g., land managers) and assist them in demonstrating the ecosystem services and socio-economic values of their conservation research.

» Provide opportunities and encourage principle investigators to compete for additional funds to deliver science.

» Use the LCC Network website and other tools as a platform for sharing key LCC services and benefits such as news, products, tools, training, science, data, documents, and open source software for conservation use or research, where appropriate.

» Disseminate science products among the Network and other interested parties, including target audiences who influence landscape and seascape conditions and resource management activities.

» Provide opportunities to train resource managers on how to apply LCC Network and others' science products to on-the-ground conservation activities.

Objective 2: Engaging Key Partners

» Capitalize on, coordinate, and target local/regional/national/international opportunities to discuss LCCs and landscape conservation at existing symposia, meetings, and other forums.

» Include key regional researchers and managers to give presentations to LCC gatherings, meetings, and workshops.

» Support, use, and fund social science approaches and human dimensions of conservation work to assess, understand, and effectively engage new partners and to assess the needs of on-the-ground users of LCC information.

» Identify and engage potential new partners across the LCC Network.

» Identify new, strategic target audiences, such as young conservation leaders, diverse audiences, development communities, planners from the built environment and others who could use science to make decisions that support conservation.

Objective 3: Communications Plan

» Identify key audiences to target outreach efforts.

» Identify priority or timely messages for targeted audiences.

» Include a specific strategic effort to target key audiences in other branches of government to increase awareness about the Network.

» Find and leverage key points of influence (i.e., LCC champions) to ensure the sustainability of the Network.

» Develop coordinated messaging from the Network to the LCC level and across the Network.

» Explain the role of LCCs in achieving lasting, sustainable landscape-scale conservation.

» Communicate LCC successes that show actual on-the-ground impact and how they have influenced management decisions through a variety of means. These successes also need to convey the complexities and challenges of achieving landscape- and seascape-scale conservation results.

» Develop appropriate tools and tacti for integrating communications acr the LCC Network.

» Open and increase availability of L products and information to diverse audiences.

Objective 4: Communications Capacity

» Build upon existing communication efforts to share learning, best practices, and identify communications needs.

» Foster regional communications communities of practice that build upon the strengths and expertise of LCCs and their individual partners.

» Improve use of existing communications tools (e.g. social media, news, and websites).

» Prioritize national LCC competitive project support funds (not individual LCC project funds) for communications support in 2015.

» Identify gaps and needs in communications support and devel a national campaign using expert public relations/marketing firms.

» Improve communications skills and effectiveness by providing training to LCC communications staff and others.

Objective 5: Sharing Lessons Learned

» Develop a Network-level means to p and distribute these lessons.

» Integrate better and more simplifie communications approaches into LCC Network gatherings, monthly LCT teleconferences, and other appropriate venues to improve the dissemination of best practices acros the Network.

DEFINITIONS

LCC NETWORK—The LCC Network is composed of the 22 individual LCCs and their active members, including the LCC Council, steering committees of all 22 LCCs, staff, partners, and other associates. The LCCs collectively form a network of resource managers, conservation practitioners, cultural communities, researchers and scientists who share a common need for scientific information, and a common interest in conservation at local, regional, and continental (or oceanic) scales. The Network fosters collaboration and partnerships among federal, provincial, state, and local governments, tribes and First Nations, indigenous peoples, non-governmental organizations, universities, and interested public and private organizations.

LCC COUNCIL—The LCC Council is a representative body of executive-level leaders from LCC partner organizations. The LCC Council supports the cooperative conservation and sustainable resource management efforts of the LCC Network, assists the LCC Network in achieving its goals, contributes to building a constituency of partners, and helps sustain the LCC initiative.

LCC COORDINATORS TEAM—The LCC Coordinators Team (LCT) is comprised of the Coordinators from each of the 22 LCCs and the LCC Network Coordinators. The LCT works on aspects of LCC Network operations and other matters as appropriate, while respecting individual LCC steering committee governance authority.

LCC SCIENCE COORDINATORS TEAM—The LCC Network Science Coordinators Team (LSCT) is comprised of all Science Coordinators from each of the 22 LCCs and the LCC Network Coordinators. The LSCT serves as a forum for communication and collaboration on technical and scientific matters among the LCCs, and between the LCC Network and other science partners.

LCC COMMUNICATIONS TEAM—The LCC Network Communications Team is comprised of all communications, engagement, and outreach staff at each of the 22 LCCs and the LCC Council. This team serves as a forum for activities in those same areas across the LCC Network.

STEERING COMMITTEES—Each LCC is governed by a voluntary steering committee, typically with representatives from conservation and resource management entities (natural and cultural). These entities include a wide variety of federal, state, territorial and international agencies; tribal and other indigenous peoples; universities; non-governmental organizations; and others located or operating within the LCC geographic region.

WWW.LCCNETWORK.ORG

Bottomland forest, Big Lake National Wildlife Refuge. JEREMY BENNETT/USFWS

Flint Hills of Kansas. EDWIN OLSON/WIKI COMMONS

Brown bear at Kodiak National Wildlife Refuge. STEVE HILLEBRAND/USFWS

LEARN MORE To find out more about how Landscape Conservation Cooperatives are working to implement this strategic plan, visit <www.lccnetwork.org/strategicplan>.

Appendix G

Goals of Individual LCCs Compared to Goals of the LCC Network Strategic Plan

The National Academy of Sciences Review Panel asked the Landscape Conservation Cooperative (LCC) staff to map the missions and goals of the individual LCCs to the goals and objectives of the LCC Network—"a mission map."

TABLE G.1 LCCs Contributing to LCC Network Goals and Objectives

LCC	1-1	1-2	1-3	1-4	1-5	2-1	2-2	2-3	2-4	2-5	2-6	3-1	3-2	3-3	4-1	4-2	4-3	4-4	4-5
Aleutian and Bering Sea Islands	X	X		X		X	X	X	X	X	X	X	X	X	X	X		X	X
Appalachian		X				X		X		X			X	X	X			X	
Arctic	X	X	X	X	X	X	X			X	X	X	X	X	X	X		X	
California	X	X	X	X	X	X	X	X	X	X	X	X	X	X	X	X	X	X	X
Caribbean	X	X	X	X	X	X	X			X		X	X	X		X		X	
Desert	X	X	X				X	X				X	X	X	X	X	X	X	X
Eastern Tallgrass Prairie and Big Rivers	X	X	X				X		X				X		X	X	X	X	
Great Basin	X		X	X			X			X		X	X		X	X	X		
Great Northern	X	X	X	X	X	X	X	X	X	X	X	X	X	X	X	X	X	X	X
Great Plains	X	X			X							X	X	X	X	X	X	X	X
Gulf Coast Prairie		X		X	X			X				X	X	X	X	X			
Gulf Coastal Plains and Ozarks	X	X	X	X	X	X	X	X				X	X	X	X	X	X	X	X
North Atlantic	X	X		X	X	X	X			X		X	X	X		X		X	
North Pacific	X	X	X			X	X		X	X		X	X	X	X	X	X	X	X
Northwest Boreal	X	X	X	X	X	X	X			X			X	X	X	X	X	X	X
Pacific Islands	X	X			X	X	X		X	X		X	X	X	X		X	X	X
Peninsular Florida		X			X	X					X	X		X	X	X	X	X	X
Plains and Prairie Potholes	X					X	X					X	X	X	X		X	X	X
South Atlantic	X	X	X	X	X	X	X			X			X	X	X	X		X	X
Southern Rockies	X	X	X	X	X	X	X			X		X	X	X	X	X	X	X	X
Upper Midwest and Great Lakes	X	X	X	X	X	X	X			X		X	X	X	X	X	X	X	
Western Alaska	X	X				X	X		X			X	X	X	X		X		

LCC	Goal 1: Conservation Strategy	Goal 2: Collaborative Conservation	Goal 3: Science	Goal 4: Communications
Aleutian and Bering Sea Islands	X	X	X	X
Appalachian	X	X	X	X
Arctic	X	X	X	X
California	X	X	X	X
Caribbean	X	X	X	X
Desert	X	X	X	X
Eastern Tallgrass Prairie and Big Rivers	X	X	X	X
Great Basin	X	X	X	X
Great Northern	X	X	X	X
Great Plains	X	X	X	X
Gulf Coast Prairie	X	X	X	X
Gulf Coastal Plains and Ozarks	X	X	X	X
North Atlantic	X	X	X	X
North Pacific	X	X	X	X
Northwest Boreal	X	X	X	X
Pacific Islands	X	X	X	X
Peninsular Florida	X	X	X	X
Plains and Prairie Potholes	X	X	X	X
South Atlantic	X	X	X	X
Southern Rockies	X	X	X	X
Upper Midwest and Great Lakes	X	X	X	X
Western Alaska	X	X	X	X

LCC Network Goal-Objective #	LCC Network Objective Description
1-1	Identify shared conservation objectives, challenges, and opportunities to inform landscape conservation at continental, LCC, island, and regional scales.
1-2	Develop then deliver (through partners) regional landscape conservation goals and designs that support resiliency and adaptation to both global change and regional landscape challenges, while ensuring the inclusion of all partners and stakeholders necessary for successful conservation.
1-3	Integrate regional or other scale-specific conservation designs to align and focus conservation action at the network scale, within available authorities.
1-4	Identify and obtain the resources required at the LCC and network scales to inform, develop, and support implementation of the conservation designs and other conservation actions.
1-5	Monitor the effectiveness of conservation design(s) and design application in terms of achieving stated outcomes, then revise as appropriate.
2-1	Create a high-functioning organizational culture for LCCs and the network.
2-2	Identify and explore opportunities for collaborative actions within the LCC Network.
2-3	Demonstrate, monitor, and evaluate the value and effectiveness of the LCC Network.
2-4	Identify new and existing training and networking opportunities for the LCC Network.
2-5	Leverage conservation planning to be opportunistic in taking advantage of current and new funding sources for conservation.
2-6	Create a network-level system for prioritizing operational needs at network and regional levels, as appropriate.
3-1	Identify shared science, information, and resource needs at the network scale.
3-2	Promote collaborative production of science and research—including human dimensions—as well as the use of experience and indigenous and traditional ecological knowledge among LCCs, Climate Science Centers (CSCs), and other interested parties; use these to inform resource management decisions, educate local communities, and address shared needs.
3-3	Demonstrate and evaluate the value and improve the effectiveness of LCC science.
4-1	Communicate the existence and application of LCC Network science, products, and tools to partners and stakeholders in a form that is understandable, publicly accessible, engaging, and relates to what matters to end users and society.
4-2	Increase two-way communication with, outreach to, and engagement of key partners across the LCC Network as well as new partners to expand the LCC Network and increase conservation impact and achievements.
4-3	Develop and implement a communications and outreach plan that identifies and uses appropriate media to clearly convey to appropriate target audiences the value and tangible successes of the LCC Network at various scales.
4-4	Build communications capacity and capabilities within the LCC Network to effectively communicate the purposes and successes of the LCC Network.
4-5	Share lessons learned across the LCC Network.

SOURCE: Provided for this report by LCC staff.

Appendix H

Committee and Staff Biographies

COMMITTEE

Dorothy J. Merritts (*Chair*) is the Harry W. and Mary B. Huffnagle Professor and chair of the Department of Earth and Environment at Franklin and Marshall College in Lancaster, Pennsylvania. In 2004–2005 she was the Flora Stone Mather Visiting Distinguished Professor at Case Western Reserve University in Cleveland, Ohio, and in 2011-2012 she was the Cox Visiting Professor at Stanford University. In the western United States, she conducted pioneering research on the San Andreas fault of coastal California, and her international work focuses on fault movements in Australia, Costa Rica, Indonesia, and South Korea. Her primary research in the eastern United States is on streams in the mid-Atlantic Piedmont, particularly in southeastern Pennsylvania and northern Maryland, where she is investigating the impact on streams of the transformation of woodlands and wetlands to a predominantly agricultural and mixed industrial-urban landscape since European settlement. She is the author of two textbooks and more than 40 scientific articles and the editor and contributing writer for numerous scientific books. Dr. Merritts has done extensive work on inquiry-based learning in the classroom, particularly for non-science majors at the undergraduate level, and has assisted in presenting original inquiry-based materials and demonstrations online through the Science Education Resource Center at Carleton College, Minnesota. Dr. Merritts received her B.Sc. in geology from Indiana University of Pennsylvania, her M.Sc. in engineering geology from Stanford University, and her Ph.D. in geology from the University of Arizona.

Brenda Barrett is the editor of the *Living Landscape Observer,* an online site that provides information and commentary on the emerging field of landscape-scale conservation, historic preservation, and sustainable communities. She served as the former Director of Recreation and Conservation at the Pennsylvania Department of Conservation and Natural Resources, directing assistance for conservation, recreation, and heritage landscape partnerships. Prior to this position she was the National Coordinator for Heritage Areas for the National Park Service in Washington, DC. Earlier in her career, she served as the Director of the Bureau for Historic Preservation at the Pennsylvania Historical and Museum Commission. Mrs. Barrett obtained an M.S. in archaeology from the University of Wisconsin and a J.D. from the Dickinson School of Law at Pennsylvania State University. She is a board member of International Council on Monuments and Sites (US/ICOMOS) and an expert member of the ICOMOS International Scientific Committee on Cultural Landscapes.

Terry Chapin is a professor emeritus of ecology at the University of Alaska, Fairbanks. His research focuses on ecosystem ecology with particular interests in the resilience of social-ecological systems and plant physiology. Dr. Chapin has been the recipient of several honors and awards. His recognitions at the University of Alaska, Fairbanks, include the Usabelli Award (for top researcher in all fields), Distinguished Professor and Distinguished Professor Emeritus recognitions, as well as a Lifetime Achievement Award. Dr. Chapin has served on several editorial and advisory boards and was president of the Ecological Society of America between 2010 and 2011. Dr. Chapin is also a member of the American Academy of Arts and Sciences and the National Academy of Sciences. He has served on several committees, boards, and roundtables at the Academies. He received a Ph.D. in biological sciences from Stanford University.

Holly Doremus is the James H. House and Hiram H. Hurd Professor of Environmental Regulation; Co-Director, Center for Law, Energy, and the Environment; and Director, Environmental Law Program at Berkeley Law. In addition to her law school teaching experience, she has taught in the graduate ecology program at the University of California, Davis; in the College of Natural Resources at the University of California, Berkeley; and at the Bren School of

Environmental Science and Management at the University of California, Santa Barbara. She has been a principal investigator on two major National Science Foundation Integrative Graduate Education and Research Traineeship Program grants and a multidisciplinary grant dealing with hydropower relicensing in California. She has co-authored papers with economists and ecologists and has been a member of two National Research Council committees. Dr. Doremus received her Ph.D. from Cornell University in plant physiology and her J.D. from the University of California, Berkeley.

Craig Groves is a Senior Scientist at The Nature Conservancy (TNC). There he currently staffs the Science for Nature and People Initiative (http://www.snap.is), which is a collaboration among TNC, the Wildlife Conservation Society (WCS), and the National Center for Ecological Analysis and Synthesis that is using a multidisciplinary team approach to addressing major conservation and science issues. Mr. Groves is also the Series Editor for the International Union for Conservation of Nature's World Commission on Protected Areas Best Practice Guidelines. Prior to working for TNC, Mr. Groves served as a conservation biologist and planner for WCS. He has published a book on conservation planning, *Drafting a Conservation Blueprint*, as well as numerous scientific articles on conservation planning and ecology. His second book, *Conservation Planning: Informed Decisions for a Healthier Planet* (with co-author Eddie Game), was published in 2015. He received an M.S. in ecology from Idaho State University.

Kenneth Haddad is the former Executive Director of the Florida Fish and Wildlife Conservation Commission (FWC). He molded a new agency (FWC) made up of multiple components from different state agencies with different cultures and different philosophies into a nationally recognized cohesive core-mission-oriented agency responsive to the needs of the 21st century. Human dimension training and techniques provided the cultural and transformational change from an isolated agency to one focusing on leadership, partnership, customer service, efficiency, and science-based decision making. Mr. Haddad was Chairman of the Science Coordinating Group of the Everglades Restoration Task Force, was a member of the Executive Committee of the Association of Fish and Wildlife Agencies, and was president of the Southeastern Association of Fish and Wildlife Agencies. He is a former Commissioner of the Atlantic States Marine Fisheries Commission and former council member of the South Atlantic Fisheries Management Council. Mr. Haddad also served as the Director of the FWC Florida Marine Research Institute, where he was responsible for applied scientific monitoring and assessment of Florida's marine resources. Prior to that, he held the position of Interim Director of the FWC Division of Marine Fisheries. Although retired, he works part time for the American Sportfishing Association (the trade association of tackle and related manufacturers, retailers, and associated industries) as their Marine Fisheries Advisor for national recreational fishing issues with focus on the Gulf of Mexico and South Atlantic. Prior to taking leadership roles, Mr. Haddad conducted applied scientific research on fisheries habitat and red tides, and specialized in remote sensing and GIS applications. He has a B.S. in biology (1974) from Presbyterian College and an M.S. in marine science (1982) from the College of Marine Science, University of South Florida.

Jessica Hellmann is an Associate Professor of Biological Sciences at the University of Notre Dame. There she also leads the Climate Change Adaptation program at the Notre Dame Environmental Change Initiative and serves as the scientific lead for Notre Dame Global Adaptation Index (ND-GAIN). In her ND-GAIN leadership role, Hellmann advises research staff and seeks the consult of Notre Dame and other experts so that ND-GAIN and its related activities capture cutting-edge knowledge about the nature of climate change and strategies for reducing the impacts of climatic change. She also directs an interdisciplinary training program for Ph.D. students, called GLOBES, that builds student capacity for scientific outreach. Dr. Hellmann's background is in ecology, where she studies the impacts of climate change on species and ecosystems and effective methods for managing nature in the face of climate change. Her research has been published in leading academic publications including the *Proceedings of the National Academy of Sciences of the United States of America*, *Conservation Biology*, and *Ecology*. She received her Ph.D. from Stanford University.

Lynn Maguire is Professor of the Practice of Environmental Decision Analysis at Duke University. Her current research uses a combination of methods from decision analysis, environmental conflict resolution, and social psychology to study environmental decision making. She focuses on collaborative decision processes where values important to the general public and stakeholders must be combined with technical analysis to determine management strategies. Her recent applications of decision analysis include the management of rare species, invasive species, and wildfire risk. Dr. Maguire is also using the principles of decision analysis to improve multicriteria rating systems, such as those used to set conservation priorities, and to develop frameworks used to value ecosystem services. She is involved in both empirical and conceptual research on perceptions of the time value of environmental resources, such as endangered species, and on the mismatches in timescales of ecological, social, and political systems.

Philip W. Mote is a Professor in the College of Earth, Ocean, and Atmospheric Sciences at Oregon State University. He also is the Director of the Oregon Climate Change Research Institute for the Oregon University System. Before joining Oregon State University, he was a Research Scientist at the

University of Washington and the State Climatologist for Washington. Dr. Mote's research interests include climate variability and change in the Pacific Northwest; regional climate modeling; mountain snowpack and its response to climate variability and change; sea level rise; impacts of climate change on water resources, forests, and shorelands; and adaptation to climate change. Among his publications in these areas is an analysis of sea level rise in the coastal waters of Washington State. Dr. Mote has served on several committees associated with climate change and sea level rise, including the National Research Council Panel on Adapting to the Impacts of Climate Change and the Intergovernmental Panel on Climate Change. He received a Ph.D. in atmospheric sciences from the University of Washington.

John O'Leary is the Assistant Director for the Massachusetts Division of Fisheries and Wildlife. He works to enhance the conservation and management of fish and wildlife resources. Mr. O'Leary also served as co-chair of the Vulnerability Assessment Sub-Committee of the Association of Fish and Wildlife Agencies Climate Change Committee, which developed a guidance document meant to aid states in making State Wildlife Action Plans climate-smart. He participated in a national working group convened by the National Wildlife Federation and the U.S. Fish and Wildlife Service that produced *Scanning the Conservation Horizon*, a guidance document centered on providing detailed information on vulnerability assessment techniques. Mr. O'Leary received an M.S. in fisheries and wildlife biology from the University of Massachusetts, Amherst.

Rebecca Rubin is the Founder, President, and CEO of Marstel-Day, LLC, an environmental consulting enterprise. She established Marstel-Day in 2002 as an expression of her commitment to the conservation of natural resources, especially habitat and open space, energy, water, and the resolution of issues at their intersections. She has extensive experience in program evaluation and policy analysis. Prior to founding Marstel-Day, she served as the Director of the Army Environmental Policy Institute and before that as a member of the professional research staff at the Institute for Defense Analyses. She was a committee member on the National Research Council Committee on Alternatives for Controlling the Release of Solid Materials from Nuclear Regulatory Commission-Licensed Facilities. Ms. Rubin has an M.A. in international security from Columbia University's School of International and Public Affairs.

Dale Strickland is the President and Senior Ecologist with Western EcoSystems Technology, Incorporated (WEST). He has more than 40 years of experience in ecological research and wildlife management. Prior to his employment with WEST, he served as a Scientist and Administrator with the Wyoming Game and Fish Department and on the faculty of the Department of Statistics at the University of Wyoming. His areas of expertise include the design and conduct of wildlife studies, impact and risk assessment, and natural resource damage assessment studies. He has taught courses in wildlife management and statistics as a visiting professor at the University of Wyoming. He contributed to documents for the National Oceanic and Atmospheric Administration regarding the quantification of injury due to oil spills. He was a committee member on the National Research Council report titled *Environmental Impacts of Wind-Energy Projects*. He served as the Executive Director of the Platte River Endangered Species Partnership. He also served as an Associate Editor and is a frequent reviewer for the *Journal of Wildlife Management*. Dr. Strickland received a Ph.D. in zoology from the University of Wyoming. He is a Certified Senior Ecologist by and a member of the Board of Certification for the Ecological Society of America and a Certified Wildlife Biologist by the Wildlife Society.

Eric Toman is an Associate Professor in the School of Environment and Natural Resources at The Ohio State University. He has an interdisciplinary background that includes training and experience in the social and natural sciences. His research focuses on developing a better understanding of the social dimensions of coupled human and natural systems. Using theory and methods from sociology and social psychology, Dr. Toman examines the factors that influence the adoption of behaviors that enable adaptation to changing environmental conditions. He received an M.S. in forest resources and a Ph.D. in forest resources from Oregon State University.

STAFF

Claudia Mengelt is a senior program officer with the Ocean Studies Board. She joined the full-time staff of the National Academies of Sciences, Engineering, and Medicine in 2005. While with the Academies, she has led several climate change studies including the *Analysis of Global Change Assessments* (2007) and *Adapting to the Impacts of Climate Change* (2010). She has also conducted several programmatic reviews such as *Strategic Guidance for the NSF's Support of Atmospheric Sciences* (2007), *Earth Observations from Space: The First 50 Years of Scientific Achievements* (2007), *Tsunami Warning and Preparedness* (2010), and the review of the new National Ocean Acidification Research Plan (2012). While at the Academies, she has also published a range of science policy articles. She obtained her M.S. in biological oceanography from the College of Oceanic and Atmospheric Sciences at Oregon State University and her Ph.D. in marine sciences from the University of California, Santa Barbara.

David Policansky received his Ph.D. in biology from the University of Oregon, where he studied evolutionary biology and ecology. He has published on life-history transitions, including the cost and timing of sexual reproduction in plants and animals; he also has published on fisheries and the interface between science and policy and on the inheritance of asymmetries in flounders. In his more than 30 years at the National Academies of Sciences, Engineering, and Medicine he has been involved in more than 35 reports, many as project director. His work has focused on management of natural resources, natural restoration, information for environmental decision making, reviews of large federal programs, and endangered species, among other topics.

Stacee Karras is an Associate Program Officer with the Ocean Studies Board. She joined the National Academies of Sciences, Engineering, and Medicine in 2012 as a fellow. She received her B.A. in marine affairs and policy with concentrations in biology and political science from the University of Miami in 2007. The following year she received an M.A. in marine affairs and policy from the University of Miami's Rosenstiel School of Marine and Atmospheric Science. Most recently, she earned her J.D. from the University of Virginia School of Law.

Heather Coleman is a Postdoctoral Fellow with the Ocean Studies Board and Board on Atmospheric Sciences and Climate. She graduated from University of California, Santa Barbara, with a Ph.D. in Environmental Science and Management after studying ecological and demographic effects of natural oil seeps on marine invertebrate populations. For her M.A. in economics she studied the history, politics, social dynamics, ecological effects, costs and benefits of restoring the Golden Horn estuary in Istanbul. She has also researched the ecological effects of marine debris, oceanic biogeochemical cycling, invasive plant ecology, and coral reef community dynamics. Before joining the Academies in 2015, Ms. Coleman aided marine conservation and resource use planning efforts in Canada and internationally as the Science and Policy Advisor for the Pacific Marine Analysis and Research Association.

Jenna Briscoe is a Senior Program Assistant with the Board on Agriculture and Natural Resources. She joined the National Academies of Sciences, Engineering, and Medicine in 2014. Previously, Ms. Briscoe worked at the University of Maryland Center for Environmental Science—Chesapeake Biological Laboratory in Solomon's Island, Maryland, where she conducted water quality testing on pre-restored and restored streams. She graduated cum laude from the University of Maryland, Baltimore County, in 2013 with a B.A. in environmental studies and a minor in sociology.